Directed Evolution Library Creation

METHODS IN MOLECULAR BIOLOGY™

John M. Walker, Series Editor

243. **Chiral Separations:** *Methods and Protocols,* edited by *Gerald Gübitz and Martin G. Schmid, 2004*
242. **Atomic Force Microscopy:** *Methods and Applications,* edited by *Pier Carlo Braga and Davide Ricci, 2004*
241. **Cell Cycle Checkpoint Control Protocols,** edited by *Howard B. Lieberman, 2004*
240. **Mammalian Artificial Chromosomes:** *Methods and Protocols,* edited by *Vittorio Sgaramella and Sandro Eridani, 2003*
239. **Cell Migration in Inflammation and Immunity:** *Methods and Protocols,* edited by *Daniele D'Ambrosio and Francesco Sinigaglia, 2003*
238. **Biopolymer Methods in Tissue Engineering,** edited by *Anthony P. Hollander and Paul V. Hatton, 2003*
237. **G Protein Signaling:** *Methods and Protocols,* edited by *Alan V. Smrcka, 2003*
236. **Plant Functional Genomics**: *Methods and Protocols*, edited by *Erich Grotewold, 2003*
235. ***E. coli* Plasmid Vectors**: *Methods and Applications*, edited by *Nicola Casali and Andrew Preston, 2003*
234. **p53 Protocols**, edited by *Sumitra Deb and Swati Palit Deb, 2003*
233. **Protein Kinase C Protocols**, edited by *Alexandra C. Newton, 2003*
232. **Protein Misfolding and Disease:** *Principles and Protocols*, edited by *Peter Bross and Niels Gregersen, 2003*
231. **Directed Evolution Library Creation**: *Methods and Protocols*, edited by *Frances H. Arnold and George Georgiou, 2003*
230. **Directed Enzyme Evolution**: *Screening and Selection Methods*, edited by *Frances H. Arnold and George Georgiou, 2003*
229. **Lentivirus Gene Engineering Protocols**, edited by *Maurizio Federico, 2003*
228. **Membrane Protein Protocols**: *Expression, Purification, and Characterization*, edited by *Barry S. Selinsky, 2003*
227. **Membrane Transporters:** *Methods and Protocols,* edited by *Qing Yan, 2003*
226. **PCR Protocols, Second Edition,** edited by *John M. S. Bartlett and David Stirling, 2003*
225. **Inflammation Protocols,** edited by *Paul G. Winyard and Derek A. Willoughby, 2003*
224. **Functional Genomics:** *Methods and Protocols,* edited by *Michael J. Brownstein and Arkady B. Khodursky, 2003*
223. **Tumor Suppressor Genes:** *Volume 2: Regulation, Function, and Medicinal Applications*, edited by *Wafik S. El-Deiry, 2003*
222. **Tumor Suppressor Genes:** *Volume 1: Pathways and Isolation Strategies,* edited by *Wafik S. El-Deiry, 2003*
221. **Generation of cDNA Libraries:** *Methods and Protocols*, edited by *Shao-Yao Ying, 2003*
220. **Cancer Cytogenetics:** *Methods and Protocols,* edited by *John Swansbury, 2003*
219. **Cardiac Cell and Gene Transfer:** *Principles, Protocols, and Applications,* edited by *Joseph M. Metzger, 2003*
218. **Cancer Cell Signaling:** *Methods and Protocols,* edited by *David M. Terrian, 2003*
217. **Neurogenetics:** *Methods and Protocols,* edited by *Nicholas T. Potter, 2003*
216. **PCR Detection of Microbial Pathogens:** *Methods and Protocols,* edited by *Konrad Sachse and Joachim Frey, 2003*
215. **Cytokines and Colony Stimulating Factors:** *Methods and Protocols,* edited by *Dieter Körholz and Wieland Kiess, 2003*
214. **Superantigen Protocols,** edited by *Teresa Krakauer, 2003*
213. **Capillary Electrophoresis of Carbohydrates,** edited by *Pierre Thibault and Susumu Honda, 2003*
212. **Single Nucleotide Polymorphisms:** *Methods and Protocols,* edited by *Pui-Yan Kwok, 2003*
211. **Protein Sequencing Protocols, 2nd ed.,** edited by *Bryan John Smith, 2003*
210. **MHC Protocols,** edited by *Stephen H. Powis and Robert W. Vaughan, 2003*
209. **Transgenic Mouse Methods and Protocols,** edited by *Marten Hofker and Jan van Deursen, 2003*
208. **Peptide Nucleic Acids:** *Methods and Protocols,* edited by *Peter E. Nielsen, 2002*
207. **Recombinant Antibodies for Cancer Therapy:** *Methods and Protocols.* edited by *Martin Welschof and Jürgen Krauss, 2002*
206. **Endothelin Protocols,** edited by *Janet J. Maguire and Anthony P. Davenport, 2002*
205. ***E. coli* Gene Expression Protocols,** edited by *Peter E. Vaillancourt, 2002*
204. **Molecular Cytogenetics:** *Protocols and Applications,* edited by *Yao-Shan Fan, 2002*
203. ***In Situ* Detection of DNA Damage:** *Methods and Protocols,* edited by *Vladimir V. Didenko, 2002*
202. **Thyroid Hormone Receptors:** *Methods and Protocols,* edited by *Aria Baniahmad, 2002*
201. **Combinatorial Library Methods and Protocols,** edited by *Lisa B. English, 2002*
200. **DNA Methylation Protocols,** edited by *Ken I. Mills and Bernie H, Ramsahoye, 2002*
199. **Liposome Methods and Protocols,** edited by *Subhash C. Basu and Manju Basu, 2002*
198. **Neural Stem Cells:** *Methods and Protocols,* edited by *Tanja Zigova, Juan R. Sanchez-Ramos, and Paul R. Sanberg, 2002*
197. **Mitochondrial DNA:** *Methods and Protocols,* edited by *William C. Copeland, 2002*
196. **Oxidants and Antioxidants:** *Ultrastructure and Molecular Biology Protocols,* edited by *Donald Armstrong, 2002*
195. **Quantitative Trait Loci:** *Methods and Protocols*, edited by *Nicola J. Camp and Angela Cox, 2002*
194. **Posttranslational Modifications of Proteins:** *Tools for Functional Proteomics,* edited by *Christoph Kannicht, 2002*
193. **RT-PCR Protocols,** edited by *Joe O'Connell, 2002*
192. **PCR Cloning Protocols, 2nd ed.,** edited by *Bing-Yuan Chen and Harry W. Janes, 2002*
191. **Telomeres and Telomerase:** *Methods and Protocols,* edited by *John A. Double and Michael J. Thompson, 2002*
190. **High Throughput Screening:** *Methods and Protocols,* edited by *William P. Janzen, 2002*
189. **GTPase Protocols:** *The RAS Superfamily,* edited by *Edward J. Manser and Thomas Leung, 2002*
188. **Epithelial Cell Culture Protocols,** edited by *Clare Wise, 2002*
187. **PCR Mutation Detection Protocols,** edited by *Bimal D. M. Theophilus and Ralph Rapley, 2002*
186. **Oxidative Stress Biomarkers and Antioxidant Protocols,** edited by *Donald Armstrong, 2002*

METHODS IN MOLECULAR BIOLOGY™

Directed Evolution Library Creation

Methods and Protocols

Edited by

Frances H. Arnold

California Institute of Technology, Pasadena, CA

and

George Georgiou

University of Texas at Austin, Austin, TX

Humana Press **Totowa, New Jersey**

Softcover reprint of the hardcover 1st edition 2003
999 Riverview Drive, Suite 208
Totowa, New Jersey 07512
humanapress.com

This publication is printed on acid-free paper. ∞
ANSI Z39.48-1984 (American Standards Institute) Permanence of Paper for Printed Library Materials.

Production Editor: Mark J. Breaugh.

Cover design by Patricia F. Cleary.

ISBN 978-1-61737-471-5 e-ISBN 978-1-59259-395-8

For additional copies, pricing for bulk purchases, and/or information about other Humana titles, contact Humana at the above address or at any of the following numbers: Tel.: 973-256-1699; Fax: 973-256-8341; E-mail: humana@humanapr.com; Website: humanapress.com

10 9 8 7 6 5 4 3 2 1

Library of Congress Cataloging in Publication Data

Preface

Biological systems are very special substrates for engineering—uniquely the products of evolution, they are easily redesigned by similar approaches. A simple algorithm of iterative cycles of diversification and selection, evolution works at all scales, from single molecules to whole ecosystems. In the little more than a decade since the first reported applications of evolutionary design to enzyme engineering, directed evolution has matured to the point where it now represents the centerpiece of industrial biocatalyst development and is being practiced by thousands of academic and industrial scientists in companies and universities around the world. The appeal of directed evolution is easy to understand: it is conceptually straightforward, it can be practiced without any special instrumentation and, most important, it frequently yields useful solutions, many of which are totally unanticipated. Directed evolution has rendered protein engineering readily accessible to a broad audience of scientists and engineers who wish to tailor a myriad of protein properties, including thermal and solvent stability, enzyme selectivity, specific activity, protease susceptibility, allosteric control of protein function, ligand binding, transcriptional activation, and solubility. Furthermore, the range of applications has expanded to the engineering of more complex functions such as those performed by multiple proteins acting in concert (in biosynthetic pathways) or as part of macromolecular complexes and biological networks.

Not surprisingly, the growth in the ranks of practitioners of directed evolution, and also in the range of new applications, has led to a proliferation of experimental methods aimed at simplifying the process and increasing its efficiency. The purpose of this and the accompanying volume in this series is to provide a compendium of experimental protocols accessible to scientists and engineers with minimal background in molecular biology.

Directed Evolution Library Creation focuses on methods for the generation of molecular diversity. Protocols for random mutagenesis of entire genes or segments of genes, for homologous and nonhomologous recombination, and for constructing libraries in vivo in bacteria and yeast are presented. Every one of these methods has been applied for directed evolution purposes. The optimal choice depends on the many factors that characterize each evolution problem, and we have often found that any of several different methods will work. Though there may be multiple molecular solutions to any given functional problem, the library made for directed evolution must nonetheless contain at least one of those solutions. And, the higher the frequency of potential solutions, the easier it is to find them. Thus, the choice of method for creating molecular diversity and its particular implementation are important. In addition to the various protocols for creating libraries, this volume also includes three chapters that describe ways to analyze libraries, particularly those made by recombination.

No directed evolution experiment is successful without a good screen or selection. *Directed Enzyme Evolution: Screening and Selection Methods* is devoted entirely to selection and screening methods that can be applied to directed evolution of enzymes. Directed evolution is not difficult, and these protocols, prepared by practitioners from many leading laboratories, should make this robust protein engineering approach accessible to anyone with a good problem.

Frances H. Arnold

George Georgiou

Contents

Contributors

Valérie Abécassis • *Centre de Génétique Moléculaire, CNRS, Gif-sur-Yvette Cedex, France*

Anna Marie Aguinaldo • *Xencor Inc., Monrovia, CA*

Miguel Alcalde • *Division of Chemistry and Chemical Engineering, California Institute of Technology, Pasadena, CA*

Frances H. Arnold • *Division of Chemistry and Chemical Engineering, California Institute of Technology, Pasadena, CA*

Geethani Bandara • *Division of Chemistry and Chemical Engineering, California Institute of Technology, Pasadena, CA*

Paul H. Bessette • *Department of Chemical Engineering, University of California at Santa Barbara, Santa Barbara, CA*

Thomas Bulter • *Department of Chemical Engineering, University of California at Los Angeles, Los Angeles, CA*

Patrick C. Cirino • *Division of Chemistry and Chemical Engineering, California Institute of Technology, Pasadena, CA*

Wayne M. Coco • *Tanox Inc., Houston, TX*

Patrick S. Daugherty • *Department of Chemical Engineering, University of California, Santa Barbara, CA*

Olga Esteban • *Department of Chemical and Biomolecular Engineering, University of Illinois, Urbana, IL*

Radu Georgescu • *Division of Chemistry and Chemical Engineering, California Institute of Technology, Pasadena, CA*

Takahiro Hohsaka • *Department of Bioscience and Biotechnology, Okayama University. Okayama, Japan*

Loïc Jaffrelo • *Centre de Génétique Moléculaire, CNRS, Gif-sur-Yvette Cedex, France*

John M. Joern • *Division of Chemistry and Chemical Engineering, California Institute of Technology, Pasadena, CA*

Lawrence A. Loeb • *Department of Biochemistry, University of Washington, Seattle, WA*

Stefan Lutz • *Department of Chemistry, Emory University, Atlanta, GA*

Kimberly M. Mayer • *Biology Department, Brookhaven National Laboratory, Upton, NY*

Peter Meinhold • *Division of Chemistry and Chemical Engineering, California Institute of Technology, Pasadena, CA*

Marco A. Mena • *Department of Chemical Engineering, University of California, Santa Barbara, CA*

Kentaro Miyazaki • *National Institute of Advanced Industrial Science and Technology (AIST), Ibaraki, Japan*
Hiroshi Murakami • *Department of Bioscience and Biotechnology, Okayama University, Okayama, Japan*
Annalee W. Nguyen • *Department of Chemical Engineering, University of California, Santa Barbara, CA*
Marc Ostermeier • *Department of Chemical and Biomolecular Engineering, Johns Hopkins University, Baltimore, MD*
Denis Pompon • *Centre de Génétique Moléculaire, CNRS, Gif-sur-Yvette Cedex, France*
Volker Schellenberger • *Genencor International Inc., Palo Alto, CA*
Olga Selifonova • *Cargill, Inc., Minneapolis, MN*
Volker Sieber • *Degussa Texturant Systems GmbH, Freising, Germany*
Jonathan J. Silberg • *Division of Chemistry and Chemical Engineering, California Institute of Technology, Pasadena, CA*
Masahiko Sisido • *Department of Bioscience and Biotechnology, Okayama University, Okayama, Japan*
Jessica L. Sneeden • *Department of Biochemistry, University of Washington, Seattle, WA*
Lianhong Sun • *Division of Chemistry and Chemical Engineering, California Institute of Technology, Pasadena, CA*
Alexander V. Tobias • *Division of Chemistry and Chemical Engineering, California Institute of Technology, Pasadena, CA*
Gilles Truan • *Centre de Génétique Moléculaire, CNRS, Gif-sur-Yvette Cedex, France*
Andrew K. Udit • *Division of Chemistry and Chemical Engineering, California Institute of Technology, Pasadena, CA*
Daisuke Umeno • *Division of Chemistry and Chemical Engineering, California Institute of Technology, Pasadena, CA*
Ryan D. Woodyer • *Department of Chemistry, University of Illinois, Urbana, IL*
Wenjuan Zha • *Department of Chemical and Biomolecular Engineering, University of Illinois, Urbana, IL*
Huimin Zhao • *Department of Chemical and Biomolecular Engineering, University of Illinois, Urbana, IL*
Tongbo Zhu • *Department of Chemical and Biomolecular Engineering, University of Illinois, Urbana, IL*

I

Mutagenesis and Recombination Methods

1

Generating Mutant Libraries Using Error-Prone PCR

Patrick C. Cirino, Kimberly M. Mayer, and Daisuke Umeno

1. Introduction

Directed evolution has become a powerful tool not only for improving the utility of enzymes in industrial processes, but also to generate variants that illuminate the relationship between enzyme sequence, structure, and function. The method most often used to generate variants with random mutations is error-prone PCR. Error-prone PCR protocols are modifications of standard PCR methods, designed to alter and enhance the natural error rate of the polymerase *(1,2)*. *Taq* polymerase *(3)* is commonly used because of its naturally high error rate, with errors biased toward AT to GC changes. However, recent protocols include the use of a newly-developed polymerase whose biases allow for increased variation in mutation type (i.e., more GC to AT changes) (*see* **Note 1**).

Error-prone PCR reactions typically contain higher concentrations of $MgCl_2$ (7 m*M*) compared to basic PCR reactions (1.5 m*M*), in order to stabilize non-complementary pairs *(4,5)*. $MnCl_2$ can also be added to increase the error-rate *(6)*. Other ways of modifying mutation rate include varying the ratios of nucleotides in the reaction *(7–9)*, or including a nucleotide analog such as 8-oxo-dGTP or dITP *(10)*. Fenton et al. *(11)* describe a mutagenic PCR protocol that uses dITP as well as provide an analysis of the effects of dITP and Mn^{2+} on PCR products. Mutation frequencies from 0.11 to 2% (1 to 20 nucleotides per 1 kb) have been achieved simply by varying the nucleotide ratio and the amount of $MnCl_2$ in the PCR reaction *(12)*. The number of genes that contain a mutation can also be modified by changing the number of effective doublings by increasing/decreasing the number of cycles or by changing the initial template concentration.

From: *Methods in Molecular Biology, vol. 231: Directed Evolution Library Creation: Methods and Protocols*
Edited by: F. H. Arnold and G. Georgiou © Humana Press Inc., Totowa, NJ

Given the same error-prone PCR conditions, two different genes will likely exhibit different mutation frequencies, primarily depending on the length and base composition of the template DNA. Thus, the best way to check the mutation frequency in an experiment is to estimate it from the fraction of inactive clones by sampling small numbers (one 96-well plate) of the error-prone PCR library. This is also a good way to test various conditions to obtain an appropriate level of mutation that allows variants with improvements to be isolated. *See* **Chapter 8** in the companion volume, *Directed Enzyme Evolution: Screening and Selection Methods*, for more detailed information on library analysis.

An expression system and high-throughput assay should be developed before a library of enzyme variants is generated. To take full advantage of the power of error-prone PCR, the assay must be accurate enough to detect small improvements and sensitive enough to detect the low levels of activity typically encountered in the beginning rounds of an evolution experiment.

2. Materials

2.1. Biological and Chemical Materials

1. Appropriate PCR amplification primers, designed to have similar melting temperatures, stored at –20°C (*see* **Note 2**).
2. Plasmid containing gene of interest to be amplified by mutagenic PCR.
3. 50X dNTP mixture: 10 m*M* each of dATP, dTTP, dCTP, and dGTP (Roche, Indianapolis, IN). Prepare 20 μL aliquots of this mixture (to avoid excessive freeze/thaw cycles) and store at –20°C.
4. Individual solutions of dNTPs (10 m*M*), stored as aliquots at –20°C (*see* **Note 3**).
5. *Taq* polymerase (Roche, Indianapolis, IN), stored at –20°C (*see* **Notes 3** and **4**).
6. 10X Normal PCR Buffer (comes with Roche *Taq* polymerase): 15 m*M* $MgCl_2$, 500 m*M* KCl, 100 m*M* Tris-HCl, pH 8.3, stored at –20°C.
7. 10X $MgCl_2$ solution: 55 m*M* prepared in water. (Sterilize before use.)
8. 1 m*M* solution of $MnCl_2$ prepared in water. (Sterilized before use.)
9. Agarose gels: 1% LE agarose in 1X TAE (40 m*M* Tris-acetate, 1 m*M* ethylene diamine tetraacetic acid [EDTA]), 0.5 μg/mL ethidium bromide.
10. PCR purification kit (Zymoclean Kit; Zymo Research, Orange, CA).
11. Appropriate restriction enzyme(s) (New England Biolabs (NEB), Beverly, MA), stored at –20°C.
12. T4 DNA ligase (Roche, Indianapolis, IN), stored at –20°C.
13. Suitable vector for expressing the mutant library: digested, gel-purified, and ready for ligation of PCR insert.
14. Competent microbial strain(s).
15. Appropriate antibiotic(s).
16. LB (Luria Broth) and LB-agar plates containing antibiotics. (Sterilize before use.)
17. Water. (Sterilize before use.)
18. High-throughput screening materials (e.g., 96-well plates for cell culture and library expression, 96-well microplates for screening, plate reader, and the like.)

2.2. Equipment

1. Microcentrifuge (Eppendorf 5417R, Brinkmann Instruments, Westbury, NY).
2. Thermocycler (Model PTC200, MJ Research, Waltham, MA).
3. Agarose gel running system.

3. Methods

3.1. Error-Prone PCR Using Taq *Polymerase*

1. Prepare purified plasmid DNA and determine its concentration (*see* **Note 5**).
2. For each PCR sample, add to tube:

   10 µL 10X normal error-prone PCR buffer,
   2 µL 50X dNTP mix,
   Additional dNTPs (optional) (*see* **Note 6**),
   10 µL 55 mM $MgCl_2$ $MnCl_2$ (optional) (*see* **Note 7**),
   30 pmol each primer,
   2 fmol template DNA (~10 ng of an 8-kb plasmid) (*see* **Note 8**),
   1 µL *Taq* polymerase (5U),
   H_2O to a final volume of 100 µL.
3. Mix sample.
4. Place tubes in thermocycler.
5. Run Error-Prone PCR Program (*see* **Note 9**):

   30 s at 94°C, 30 s at annealing temperature for primers (*see* **Note 10**),
   1 min at 72°C (for a ~1 kb gene) (*see* **Note 11**),
   14–20 cycles (*see* **Note 12**),
   5–10 min at 72°C final extension,
   4°C (to protect samples overnight if necessary).
6. Run a sample of the product on a gel to estimate the yield of full-length gene.
7. Purify PCR products either by gel electrophoresis (removes plasmid DNA) or by Zymoclean Kit (*see* **Note 13**).
8. Digest with appropriate restriction enzymes (*see* **Note 2**). Clean the digested insert, ligate into expression vector, and transform the mutant library into appropriate host strain (*see* **Note 14**).
9. Grow cultures expressing the mutant library (e.g., in 96-well format) and perform the corresponding enzyme activity assay.
10. Determine from the activity profiles of the expressed mutant libraries the most suitable error conditions for screening, and continue screening that library (*see* **Note 15**). *See* **Fig. 1** for example activity profiles from mutant libraries prepared under various mutagenic conditions. In general, it is desirable to obtain mutants that contain only a single amino-acid substitution compared to the parent sequence. Higher mutation rates make it difficult to distinguish beneficial point mutations from those that are neutral or even slightly deleterious. Additionally, the fraction of mutants with improved function decreases as the mutation rate is increased. Thus, an appropriate PCR error rate for directed evolution corresponds to a mutation frequency of ~2 to 5 base substitutions per gene.

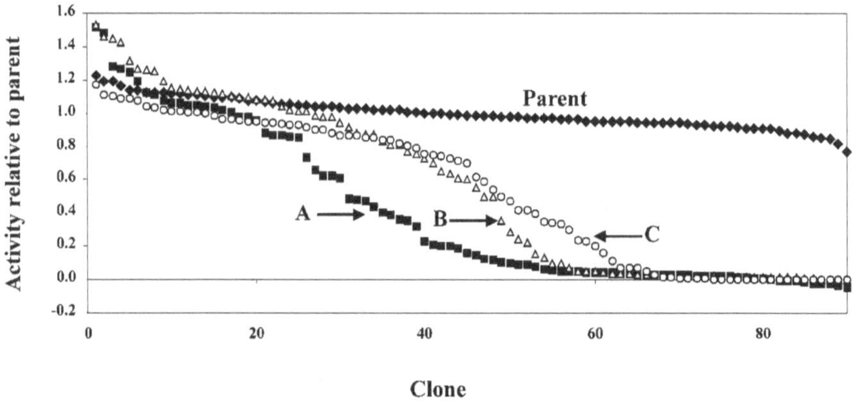

Fig. 1. Activity profiles for libraries made under different mutagenic PCR conditions. Activities are reported relative to the average activity of the parent enzyme used to prepare that generation and are plotted in descending order. The parent gene is 1.4 kb and codes for the heme domain of cytochrome P450 BM-3. The plot labeled **Parent** (◆) represents parent enzyme activity measured across an entire 96-well plate. The standard deviation in parent activity is 9.2%. The remaining three plots depict the activity profiles from 96-well plates containing different mutant libraries. All three error-prone PCR reactions contained 20 fmole of the parent gene as template, plus 7 m*M* $MgCl_2$, 0.2 m*M* each of dGTP and dATP, and 1.0 m*M* each of dCTP and dTTP. Additionally, reaction **A** (■) contained 0.1 m*M* $MnCl_2$, reaction **B** (△) contained 0.05 m*M* $MnCl_2$, and reaction **C** (○) contained no $MnCl_2$. Libraries **A**, **B**, and **C**, respectively, consist of 45%, 40%, and 31% mutants with less than 10% of the parent enzyme's activity.

Typically an error rate resulting in a library with 30–40% of mutants having less than 10% of the parent enzyme's activity (i.e., "dead" mutants) is suitable, although this value will vary depending on the enzyme and the function assayed.

4. Notes

1. Stratagene's Genemorph kit, which includes its own error-prone PCR protocol, uses a polymerase ("Mutazyme") that exhibits a mutation bias quite different from that of *Taq* polymerase. Whereas *Taq* polymerase preferentially introduces AT to GC mutations, Mutazyme mutations are biased toward GC to AT changes. It may be desirable to combine the mutation biases of these polymerases by alternating between them in successive generations, or by creating separate mutant libraries using both polymerases in a single generation.
2. Error-Prone PCR primers can be designed to anneal outside the restriction sites that will be used for subcloning or can be designed to include the restriction sites as part of the primer sequence. In our experience, higher levels of ligation efficiency are obtained when primers are located far outside the restriction sites, presumably because of better digestion efficiency.

3. dNTPs may be present in either equimolar or unbalanced amounts. An unbalanced mixture promotes misincorporation and helps to reduce the natural error bias of *Taq* polymerase. Several variations have been used, including increasing the amount of dGTP ***(13)***, increasing both dCTP and dTTP ***(2)***, or increasing all but dATP ***(1)***. The concentration of certain nucleotides can also be decreased ***(8)***, or the total nucleotide concentration per reaction can be decreased ***(6)***. In our experience, a suitable unbalanced dNTP mixture includes, as final concentrations, 0.2 m*M* each of dGTP and dATP and 1.0 m*M* each of dCTP and dTTP. This can be prepared using the 50X mixture of dNTPs (10 m*M* each) followed by addition of individual solutions of 10 m*M* dCTP and dTTP (stock solutions are available from Roche).
4. *Taq* polymerase from various sources may affect both PCR yield as well as error rate. For example, we have obtained quite different library profiles using *Taq* from Promega (Madison, WI) as opposed to *Taq* from Boehringer-Mannheim (Indianapolis, IN).
5. The concentration of the DNA template should be estimated by comparing its intensity on an ethidium-bromide stained gel to the intensity of the bands in a commercially-prepared DNA marker whose concentrations are defined by the manufacturer.
6. The most often used variable for adjusting the error-rate is the concentration of $MnCl_2$ added to the PCR reaction. Often error-rates are sufficient without $MnCl_2$ due to the natural error of the polymerase, the increased Mg^{2+} concentration, the unbalanced dNTPs (*see* **Note 3**), and the low amount of template added. By varying the amount of $MnCl_2$ added to the reaction, the mutation rate can be varied from ~1–5 nucleotides per 1 kb ***(12)***. The $MnCl_2$ should be added to the reaction last to prevent precipitation ***(2)*** and the amount of water in the sample should be adjusted so that the total volume remains 100 µL. *See* **Note 7** for more information.
7. It is a good idea to prepare several different error-prone libraries with varying concentrations of $MnCl_2$ (e.g., 0.0, 0.01, 0.05, 0.10, and 0.15 m*M*) in the PCR reaction. A master mix is useful and is prepared by mixing into one tube sufficient amounts of each component except the variable ($MnCl_2$) and then aliquoting into separate tubes. Different amounts of $MnCl_2$ (from a 1 m*M* stock, for example) are then added to each tube and the samples are brought to their final volume with water.
8. Up to 20 fmol template can be used in a reaction. Template concentration is another variable that influences error rate, with lower template concentrations resulting in higher error rates.
9. If the thermocycler does not have a heated lid, add a drop of sterile mineral oil to the top of each sample to prevent evaporation during cycling.
10. Annealing temperature will vary given the length and composition of the PCR primers. If the melting temperature is not provided by the oligonucleotide manufacturer, a good rule-of-thumb is 4 × GC + 2 × AT – 5 = annealing temperature. Note that both primers should have similar melting temperatures.
11. Increase the extension time for longer genes (~1 min per 1 kb).

12. The number of cycles can be increased to increase the number of genes that contain mutations.
13. Purification by gel electrophoresis is recommended. If the PCR products are not purified by gel electrophoresis, the PCR reaction should be digested with *Dpn*I (cuts only methylated DNA) to eliminate template DNA prior to using a PCR cleanup kit. In addition, *Taq* polymerase binds tightly to the ends of PCR products and cannot be easily eliminated using silica-based cleanup kits ***(14)***. *Taq* can interfere with ligation by filling in the ends of digested DNA and lowering the number of transformants obtained. To eliminate *Taq* polymerase after the PCR reaction, one protocol suggests adding EDTA to 5 m*M*, sodium dodecyl sulfate (SDS) to 0.5%, and proteinase K to 50 μg/mL, and incubating at 65°C for 15 min prior to using a PCR cleanup kit ***(14)***.
14. It is important to quantify the background level of transformants obtained from a ligation of the vector alone as a control. If this level is high it will influence the shape of the mutant library activity landscape. To minimize this background, the vector can be treated with shrimp alkaline phosphatase (SAP) prior to ligation, which prevents the vector from ligating to itself without an insert.
15. The activity profiles for several libraries of varying error rate can be estimated from one or two 96-well plate assays from each library, and the best library can be chosen for further screening. **Figure 1** shows activity profiles from libraries prepared under varying mutagenic PCR conditions.

References

1. Leung, D. W., Chen, E., and Goeddel, D. V. (1989) A method for random mutagenesis of a defined DNA segment using a modified polymerase chain reaction. *Technique* **1,** 11–15.
2. Cadwell, R. C. and Joyce, G. F. (1992) Randomization of genes by PCR mutagenesis. *PCR Methods Appl.* **2,** 28–33.
3. Keohavong, P. and Thilly, W. G. (1989) Fidelity of DNA polymerases in DNA amplification. *Proc. Natl. Acad. Sci. USA* **86,** 9253–9257.
4. Ling, L. L., Keohavong, P., Dias, C., and Thilly, W. G. (1991) Optimization of the polymerase chain reaction with regard to fidelity: modified T7, Taq, and vent DNA polymerases. *PCR Methods Appl.* **1,** 63–69.
5. Cadwell, R. C. and Joyce, G. F. (1994) Mutagenic PCR. *PCR Methods Appl.* **3,** S136–S140.
6. Lin-Goerke, J. L., Robbins, D. J., and Burczak, J. D. (1997) PCR-based random mutagenesis using manganese and reduced dNTP concentration. *Biotechniques* **23,** 409–412.
7. Fromant, M., Blanquet, S., and Plateau, P. (1995) Direct random mutagenesis of gene-sized DNA fragments using polymerase chain reaction. *Anal. Biochem.* **224,** 347–353.
8. Nishida, Y. and Imanaka, T. (1994) Alternation of substrate specificity and optimum pH of sarcosine oxidase by random and site-directed mutagenesis. *Appl. Environ. Microbiol.* **60,** 4213–4215.

9. Shafikhani, S., Siegel, R. A., Ferrari, E., and Schellenberger, V. (1997) Generation of large libraries of random mutants in Bacillus subtilis by PCR-based plasmid multimerization. *Biotechniques* **23,** 304–310.
10. Spee, J. H., de Vos, W. M., and Kuiper, O. P. (1993) Efficient random mutagenesis method with adjustable mutation frequency by use of PCR and dITP. *Nucl. Acids Res.* **21,** 777–778.
11. Fenton, C., Hao, X., Petersen, E. I., Petersen, S. B., and El-Gewely, M. R. (2002) Random mutagenesis for protein breeding, in *In Vitro Mutagenesis Protocols*, 2nd Ed. (Braman, J., ed.), Humana Press, Totowa, NJ, pp. 231–241.
12. Zhao, H., Moore, J. C., Volkov, A. A., and Arnold, F. H. (1999) Methods for optimizing industrial enzymes by directed evolution, in *Manual of Industrial Microbiology and Biotechnology* ASM, Washington, DC, pp. 597–604.
13. Clontech, Palo Alto, CA (1999) Diversify™ PCR Random Mutagenesis Kit. *CLONTECHniques* **14,** 14–15.
14. Wybranietz, W. A. and Lauer, U. (1998) Distinct combination of purification methods dramatically improves choesive-end subcloning of PCR products. *Biotechniques* **24,** 578–580.

2

Preparing Libraries in *Escherichia coli*

Alexander V. Tobias

1. Introduction

The process of preparing libraries of mutagenized or recombined gene sequences for screening or selection in *Escherichia coli* is a special application of cohesive-end subcloning *(1)*. PCR products are digested with restriction endonucleases, ligated into an expression vector digested with the same enzymes, and the resultant recombinant plasmids are transformed into supercompetent bacteria. One difference between routine subcloning of a single gene and preparing libraries is that in the latter case, there can be little allowance for the presence of transformants containing recircularized vector and no insert (so-called "background" ligation products). Furthermore, ligation and transformation of the recombinant plasmids must be performed using materials and conditions that yield a sufficient number of transformants (~10^3–10^5) for identifying variants exhibiting desired properties.

Background transformants are a nuisance in routine subcloning, but do not generally ruin the experiment; one can merely pick a few transformants for growth and test them for the presence of the insert by hybridization, PCR, sequencing, or restriction digest. One does not have this luxury when preparing libraries since the number of clones to be screened vastly exceeds the number that can be tested for the presence of insert. Background transformants waste screening effort and decrease the diversity of the transformant library. There can be zero tolerance for background transformants when one is screening for loss-of-function mutants, as these are sure to be confused with desired clones. When screening a library for gain-of-function mutants, a small fraction (<1%) of background transformants may be acceptable, since these will not be confused with positive clones.

From: *Methods in Molecular Biology, vol. 231: Directed Evolution Library Creation: Methods and Protocols*
Edited by: F. H. Arnold and G. Georgiou © Humana Press Inc., Totowa, NJ

Probably the most difficult and frustrating aspect of preparing libraries in *E. coli* is the requirement for a high number of transformants per plate compared to routine subcloning. For example, to screen 50,000 clones, one can grow 50 plates of 1000 clones each or 500 plates of 100 clones each. Obviously, the latter case is much more labor- and resource-intensive, making the former case vastly preferred. The major obstacle to obtaining a sufficient number of transformants per plate lies in the poor transformation efficiency of ligation products—typically one to three orders of magnitude lower than that of supercoiled plasmids. This can represent a challenge to the researcher who wants to screen 10^4–10^5 clones. Nonetheless, with careful preparation of DNA fragments and some optimization of the ligation reaction, one can obtain transformation efficiencies of 10^7 colony-forming units (cfu) per microgram of vector DNA when transforming supercompetent cells. Generally, a transformation efficiency of ~10^6 cfu/μg DNA provides a sufficient number of transformants to screen for most directed evolution applications.

2. Materials

1. Mutagenic or recombination PCR product.
2. Plasmid vector for cloning.
3. Sterile distilled water (dH_2O).
4. DpnI endonuclease (New England Biolabs, Beverly, MA).
5. Restriction endonucleases and buffers.
6. Zymo-5 DNA Clean and Concentrator Kit (Zymo Research, Orange, CA).
7. Zymo-25 DNA Clean and Concentrator Kit (Zymo Research, Orange, CA).
8. Agarose-dissolving buffer (Zymo Research, Orange, CA).
9. 1% agarose gels (analytical and preparative).
10. Gel electrophoresis equipment.
11. UV transilluminator.
12. Sterile razor blades.
13. Shrimp alkaline phosphatase and 10X buffer (USB, Cleveland, OH).
14. T4 DNA ligase and 10X buffer (Roche, Indianapolis, IN) (*see* **Note 1**).
15. Control circular plasmid.
16. Transformation-competent *E. coli* cells.
17. SOC media (*see* **Note 2**).
18. Luria-Bertani agar plates supplemented with appropriate antibiotic.

3. Methods

3.1. Choice of Restriction Sites, PCR Primers, and Vector

Two different restriction sites must be used for subcloning an insert library in order to ensure proper insert orientation in the recombinant plasmids. Avoid restriction endonucleases known by the manufacturer to have "star" activity (ability to cleave at sequences that are similar but not identical to the primary recognition sequence). If the two restriction endonucleases have compatible buffers and operating temperatures, it may be possible to save time and eliminate purification steps by digesting with both endonucleases simultaneously. This should be verified experimentally, even if the supplier's instructions indicate that the two enzymes are compatible and may be used simultaneously.

Do not design PCR primers with the restriction site at the 5' terminus. Rather, primers should have at least five "spacer" nucleotides 5' of the restriction site. This permits more efficient digestion of the PCR product by restriction endonucleases. The spacer nucleotides should be complementary to the PCR template. The extra annealing this provides may improve the PCR reaction.

If possible, choose a vector with an ampicillin- or kanamycin-resistance gene. Compared to chloramphenicol, these two antibiotics place less metabolic burden on growing colonies. Use of chloramphenicol can result in growth delays and fewer surviving colonies.

Optimizing ligation and transformation efficiencies may involve trying different cloning vectors and restriction endonuclease recognition sites. The researcher is encouraged to attempt the protocols in this chapter with several different vectors and restriction sites in order to increase the chances of obtaining adequate-sized libraries of transformants for screening.

3.2. Preparation of Insert Library

1. Run a 1 μL aliquot of the completed PCR reaction on an agarose gel to estimate the concentration of PCR product.
2. Digest the PCR template DNA by adding 1 μL (20 U) *Dpn*I directly to the entire completed PCR reaction (*see* **Note 3**).
3. Incubate at 37°C for 1 h.
4. Purify 2 μg of PCR product using the Zymo-5 DNA Clean and Concentrator kit. Follow the instructions supplied with the kit. Elute the DNA from the Zymo-5 column with 20 μL dH_2O (*see* **Note 4**).
5. Digest the purified PCR product with two units of each restriction endonuclease (*see* **Note 5**), following the guidelines supplied with the enzymes. The total volume of the insert digestion reaction(s) should be 100 μL.
6. Purify the digested insert using the Zymo-5 Kit. Elute the DNA from the Zymo-5 column with 20 μL dH_2O.
7. Run a 1-μL aliquot of the insert DNA on an agarose gel to estimate its concentration.

3.3. Preparation of Plasmid Vector

1. Purify 2 μg of vector DNA with the Zymo-5 kit (*see* **Note 6**). Elute the DNA from the Zymo-5 column with 80 μL dH_2O.
2. Digest the vector DNA with the two units of each restriction endonuclease (*see* **Note 5**), following the guidelines supplied with the enzymes. The total volume of the vector digestion reaction(s) should be 100 μL.

3. Run the digested vector DNA on a preparative agarose gel.
4. Viewing the gel with an UV transilluminator under preparative illumination intensity, excise the band corresponding to doubly cut vector DNA with a sterile razor blade.
5. Place the excised band in a 2 mL centrifuge tube and fill to the top with Agarose-dissolving buffer. Incubate at 50°C until the agarose is completely dissolved (~10 min).
6. Pass the entire amount of agarose-dissolving buffer containing the vector DNA through a single Zymo-25 column in order to bind all the DNA to the column (this will require several runs on the centrifuge since the volume of a column is only about 0.6 mL).
7. Wash the vector DNA bound to the column as described in the Zymo-25 DNA Clean and Concentrator Kit instructions.
8. Elute the vector DNA with 80 μL dH_2O.
9. Run a 2-μL aliquot of the doubly cut vector DNA on an agarose gel to estimate its concentration.
10. Set aside a 100 ng aliquot of purified, linearized vector DNA for use in control ligation A (*see* **Subheading 3.4.**).
11. Combine the remaining doubly cut vector DNA, 10 μL 10X phosphatase buffer, 1 μL (1 U) phosphatase, and dH_2O to 100 μL. Incubate the phosphatase reaction at 37°C for 1 h (*see* **Note 7**).
12. Purify the phosphatase-treated vector DNA using the Zymo-5 kit. Elute the DNA with 20 μL dH_2O.
13. Run a 1-μL aliquot of the purified phosphatase-treated vector DNA on an agarose gel to estimate its concentration.

3.4. Ligation of Insert and Vector DNA

Obtaining a sufficient number transformants for screening in directed evolution requires obtaining a near-maximal yield of desired recombinant plasmids from the ligation reaction. To achieve this, some optimization of the reaction is usually necessary. One of the most important parameters in a ligation reaction is the ratio of insert to vector molecules. Maximum yield of desired ligation products is usually achieved when this ratio is approximately 2:1 *(2)*. Nevertheless, running several parallel ligation reactions at different values of this parameter (ligations C-E below) should increase the chances of obtaining near-maximum yield in one of the reactions.

Two different control ligation reactions lacking insert DNA should be performed. Ligation A is carried out in order to estimate the amount of background ligation products present in ligations C-E. Ligation B is performed to evaluate (by comparison to Ligation A) the effectiveness of phosphatase treatment.

1. Set up the following ligation reactions: ligation A: 100 ng phosphatase-treated vector (control for vector recircularization); ligation B: 100 ng untreated vector

(control for phosphatase activity); ligation C: 100 ng phosphatase-treated vector + equimolar amount of insert DNA; ligation D: 100 ng phosphatase-treated vector + 2-fold molar excess of insert DNA; ligation E: 100 ng phosphatase-treated vector + 3-fold molar excess of insert DNA.
2. To each ligation, add 2 μL 10X ligase buffer, 1 μL (1 U) ligase, and dH_2O to 20 μL.
3. Incubate at 16°C for 12 h (*see* **Note 8**).

3.5. Transformation of Ligations

Transformation-ready competent cells are available from suppliers such as Stratagene (La Jolla, CA). Alternatively, Zymo Research supplies an excellent kit for making one's own competent *E. coli* cells.

1. Transform Ligations A-E, into the cells, following the transformation protocol supplied with the competent cells or competent cell kit (*see* **Note 9**). As a control for the efficacy of the antibiotic, transform an equivalent amount of dH_2O to the same amount of cells. As a control for the efficiency of transformation, transform an equivalent amount (in ng) of control plasmid to the same amount of cells.
2. Spread the transformed cells onto the agar plates as instructed in the transformation protocol supplied with the competent cells or kit.
3. Incubate the plates overnight at 37°C.

4. Notes

1. The 10X ligase buffer contains ATP, which is unstable. To prevent degradation by repetitive thawing and freezing, this buffer should be divided into 10–20 μL aliquots for single use and stored at –20°C.
2. SOC media: to make 1 L, add 20 g bacto tryptone, 5 g yeast extract, 0.5 g NaCl, and 10 ml KCl (250 m*M*) to 975 mL distilled water. Autoclave for 30 min on liquid cycle. Immediately before use, add 5 mL $MgCl_2$ solution (2 *M*) and 20 mL glucose solution (1 *M*). Both solutions must be sterilized by filtration. Store refrigerated.
3. *Dpn*I is a restriction endonuclease with a four-base recognition site that digests only methylated DNA. It is used here to destroy plasmid template DNA, so that it does not contaminate the recombinant plasmid library. *Dpn*I treatment is not necessary if the PCR template is not a circular plasmid capable of replicating in *E. coli*, and will not destroy plasmid DNA purified from methylation negative bacteria.
4. Eluting with sterile distilled water warmed to 55°C increases recovery of DNA from Zymo columns.
5. If digesting with one endonuclease at a time, purify the DNA using the Zymo-5 kit after the first digestion, then digest with the other endonuclease.
6. It is preferable to begin with a vector containing an insert at the site where the library is to be ligated, rather than a vector containing only a small spacer sequence between the two cloning restriction sites. In the former case, "empty," doubly cut vector DNA is easily and efficiently isolated by preparative gel electrophoresis. In the latter case, the separation between uncut, singly cut, and dou-

bly cut vector DNA by preparative gel electrophoresis is less efficient. This can result in significant contamination of the recombinant plasmid library by uncut vector DNA.

7. Phosphatase treatment removes the 5'-phosphate groups from the linearized vector DNA, preventing recircularization of vector DNA during the ligation reaction. The insert DNA, which still possesses 5'-phosphate groups, will efficiently ligate with the phosphatase-treated vector, forming a circular DNA molecule with two nicks *(2)*. Because circular DNA, even if nicked, transforms much more efficiently than linear DNA *(3)*, nearly all of the transformants should harbor recombinant plasmids.
8. The low temperature enhances the stability of base pairing between the complementary cohesive ends of the insert and vector. This favors the desired intermolecular ligation reaction over vector recircularization. Incubating ligations at 16°C for more than 12 h will deleteriously affect the reactions.
9. Optimal transformation efficiencies are usually attained by mixing 1 μL of each ligation with 50 μL cells. If the protocol supplied with the cells includes an incubation step with SOC media, it is recommended that this step be performed for a maximum of 20 min. This reduces the likelihood of cell doubling, which reduces the occurrence of transformants harboring identical recombinant plasmids.

References

1. Delidow, B. C. (1993) Molecular Cloning of Polymerase Chain Reaction Fragments with Cohesive Ends, in *PCR Protocols - Current Methods and Applications* (White, B. A., ed.), Humana Press, Totowa, NJ, pp. 217–228.
2. Sambrook, J., Fritsch, E. F., and Maniatis, T. (1989) *Molecular Cloning: A Laboratory Manual.* 2nd ed, Cold Spring Harbor Laboratory Press, Cold Spring Harbor, NY.
3. Conley, E. C. and Saunders, J. R. (1984) Recombination-dependent recircularization of linearized pBR322 plasmid DNA following transformation of *Escherichia coli. Mol. Gen. Genet.* **194,** 211–218.

3

Preparing Libraries in *Saccharomyces cerevisiae*

Thomas Bulter and Miguel Alcalde

1. Introduction

High recombination frequency and ease of manipulation made the budding yeast, *Saccharomyces cerevisiae*, the model eukaryotic organism for studies on homologous recombination ***(1,2)***. Mutagenesis of intrinsic genes ***(3)*** and the construction of vectors ***(4)*** are applications of in vivo recombination in molecular biology. Next to *Escherichia coli*, *S. cerevisiae* is the most commonly used host organism in directed evolution ***(5–9)***. Its well developed recombination apparatus facilitates mutant library construction. Ligation of mutant genes into expression vectors is in many cases a tedious and non-robust step that needs fine tuning for new plasmid-gene combinations. Yeast gap repair can substitute for ligation to give more reliable high transformation frequency and shortening the protocol for library expression. In gap repair, the mutant gene inserts are cotransformed with open plasmid that contains sequences homologous to the ends of the inserts on both ends. Homologous recombination combines these to form complete plasmids.

Homologous recombination is a valuable tool in creating highly diverse libraries from members of gene families ***(10)***. It is frequently used to recombine mutations created in mutagenesis experiments to find the best combination of beneficial mutations and to eliminate neutral or deleterious sequence changes. Highly efficient in vitro recombination methods have been developed ***(11,12)***. One advantage of in vivo recombination is that during recombination the proofreading apparatus of the yeast cell prevents the appearance of additional mutations that are common for in vitro methods. The low recombination efficiency in *E. coli* effectively limits in vivo shuffling to genes expressed in yeast ***(13)***. In yeast, in vivo recombination is an easy, fast, and non-mutagenic method.

From: *Methods in Molecular Biology, vol. 231: Directed Evolution Library Creation: Methods and Protocols*
Edited by: F. H. Arnold and G. Georgiou © Humana Press Inc., Totowa, NJ

In many cases, the beneficial mutations found in a directed evolution experiment can be accumulated for futher improvements. If a low mutation frequency is used, the improvement achieved in the targeted trait can be assigned to specific mutations. With only a few mutations to combine, site-directed mutagenesis is often more efficient than random recombination. Furthermore, the closer the mutations are on the gene, the less likely is their independent recombination with homology-based methods. Deleterious mutations that are close to beneficial ones can also escape elimination by homologous recombination. In such cases, it is best to use site-directed recombination with primers synthesized for the mutation sites with 50% wildtype sequence. In this way, a library that includes all possible combinations of the mutations can be constructed. In *E. coli* site-directed mutagenesis is commonly done by assembly PCR. In yeast, this process can be simplified because gap repair can replace the second PCR reaction as well as the ligation step.

In this chapter, we describe protocols for in vivo recombination and in vivo assembly recombination in yeast. These methods have been used to recombine mutant genes that were generated during the directed evolution of a fungal laccase in *S. cerevisiae* ***(14)***.

2. Materials and Equipment

2.1. Materials

2.1.1. Chemicals

All chemicals used were reagent grade purity.

1. dNTPs (Boehringer-Mannheim, Indianapolis, IN).
2. Appropriate PCR primers.
3. Agarose.
4. PCR purification: QIAquick PCR purification Kit (Qiagen Inc., Valencia, CA).
5. DNA extraction from agarose gels: QIAEX II (Qiagen Inc., Valencia, CA).
6. Yeast Transformation: the Gietz Lab Transformation Kit (Tetra Link, Amherst, NY).

2.1.2. Biological Materials

1. *E. coli* competent cells (Stratagene, La Jolla, CA).
2. *S. cerevisiae*, e.g., Protease deficient *S. cerevisiae* strain BJ 5465 (ATCC 208289).
3. Expression shuttle vector containing the gene of interest under appropriate promoter, a signal sequence for secretion and selection markers for *S. cerevisiae* and *E. coli*. For example: pJRoC30, Gal10 promoter, *Myceliophthora thermophila* laccase gene with native signal sequence, markers uracil and ampicillin.
4. Gene variants from different hosts or created by random mutagenesis.
5. Restriction endonucleases.
6. Proofreading polymerase, e.g. pfu (Stratagene, San Diego, CA).

2.1.3. Buffers and Solutions

1. Chloramphenicol stock solution: 25 mg chloramphenicol in 1 mL of ethanol.
2. SC-drop-out plates:* 6.7 g yeast nitrogen base, 50 mg L-His, 50 mg L-Trp, 50 mg L-Leu, 50 mg adenine hemisulfate, 15 g bacto agar, 100 mL 20 % filtered glucose,** 1000 μL chloramphenicol stock solution** + dd H_2O to 1000 mL.
3. TAE-buffer (50X): 121 g Tris-base; 28.05 mL glacial acetic acid, 50 mL 0.5 *M* ethylenediaminetetraacetic acid (EDTA) pH 8.0 + ddH_2O to 500 mL.

2.2. Equipment

1. Agarose gel electrophoresis system.
2. Spectrophotometer (Model BioSpec-1601, Shimadzu). Software UVProbe Version 1.01.
3. Thermocycler (Model PTC200, MJ Research, Waltham, MA).
4. Yeast Transformation: the Gietz Lab Transformation Kit (Tetra Link, Amherst, NY).
5. Glass beads, 6 mm.

3. Methods

3.1. In Vivo Recombination in Yeast

1. Digest the plasmid for recombination with appropriate restriction endonucleases (*see* **Note 1**).
2. Purify the opened plasmid by gel purification (QIAEX II, Qiagen). Measure the absorption of the preparation at 260 nm to assess its concentration (*see* **Note 2**).
3. Amplify the mutant genes that will be recombined from the plasmids by PCR (*see* **Note 3**).
4. Gel purify the PCR products and measure the concentrations of the purified products.
5. Prepare an equimolar mixture of the mutant genes that will be recombined.
6. Mix the mutant genes with the preparation of the open vector in a molar ratio of 5:1 with not less than 200 ng open plasmid per 50 μL cell suspension. Example: 200 ng open plasmid (10 kb), 200 ng mutant gene mixture (2 kb).
7. Transform the mixture into *S. cerevisiae* following the protocol of the Gietz yeast transformation kit (*see* **Note 4**).
8. Plate an appropriate amount of the transformation mix onto SC-Drop out plates using glass beads and incubate at 30°C for 3–4 d (*see* **Note 5**).

3.2. In Vivo Assembly Recombination

1. Choose the mutations that will be recombined so that the distance between them is either smaller than 15 bp so that they can be recombined within one primer or higher than 150 bp so that PCR products of at least 170 bp are created (*see* **Note 6**).
2. Choose the mutant with the most mutations as the parent for the recombination.

* Store in darkness (light sensitive).

** Added after autoclaving.

3. Synthesize a pair of sense and antisense primers for every mutation site. If the mutation is going to be evaluated include 50% wild type and 50% mutated sequence so that it will be reverted if deleterious. Prepare also two external non-mutagenic primers (20–30 bp) that bind within the plasmid in at least 150 bp distance to the first mutagenic primer (*see* **Note 7**).
4. Carry out PCR reactions (One more than the number of mutagenic primers synthesized). Use proofreading polymerase, non-mutagenic conditions and a low amount of template DNA (~0.1 ng/μL) (*see* **Note 8**).
5. Purify the PCR reactions using a column (Zymoclean, Zymo Research).
6. Gel purify the concentrated samples (QIAEX II, Qiagen) (*see* **Note 9**).
7. Prepare a mixture of the purified PCR products with a minimum amount of 10 ng per fragment (*see* **Note 10**). Add the open vector (200 ng).
8. Transform the mixture into *S. cerevisiae* following the protocol of the Gietz yeast transformation kit (*see* **Note 4**).
9. Plate an appropriate amount of the transformation mix onto SC-Drop out plates using glass beads and incubate at 30°C for 3–4 d (*see* **Note 5**).

4. Notes

1. The choice of the endonucleases dictates the parts of the gene and the plasmid that will be included in the recombination. In order to exclude parts of the gene from recombination (e.g., signal sequences, specific domains), the restriction has to be done with the plasmid including the wild type gene of interest, and a restriction site has to be created at the end of the excluded sequence by changing the codons of the amino acids at the site.

   The restriction sites and the positions of the primers used for the amplification of the genes regulate the length of the homologous sequences in open plasmid and genes. This overhang length influences the recombination frequency between gene and open plasmid and therefore the transformation efficiency. Transformation efficiency does not change much if the homologous sequences are longer than 50 bp. If the overhangs are smaller, efficiency is compromised. For creation of libraries for directed evolution, 20–50 bp homology is good for making libraries of ~10,000 clones per transformation (Gietz kit yeast transformation kit, 50 μL cells, 1 μg DNA). Long stretches of wildtype or mutant DNA in the open plasmid should be avoided because this produces a library that is biased towards the sequence in the open plasmid.
2. Concentrations can also be estimated from the gel, but an accurate measurement allows the exact adjustment of the relative amounts of the ingredients of the DNA mix and therefore helps to prevent biases in the recombination.
3. New mutations can be introduced within the shuffling experiment by using error-prone PCR conditions for amplification of the parent genes. If desired, mutations can be avoided completely by using proofreading polymerase and non-mutagenic conditions for the PCR.
4. Transformation by the lithium acetate method is described elsewhere *(15)*. The necessary solutions can be prepared as described there or can be purchased from

Tetra Link (Gietz yeast transformation kit). Alternatively, electroporation can be used for the transformation of the library ***(16)***.

The Gietz kit loses efficiency gradually upon storage. Evaporation of the PEG might be the reason for the instability. If the kit has not been used for more than 2 mo, the transformation efficiency should be tested before transforming a library to assess the amount of transformation mix that has to be plated in order to get the right density of colonies for the picking robot.

5. Like library creation, library screening protocols depend on the host organism used. Differences in physiology and growth characteristics between bacteria and yeast necessitate variations in the screening procedures (*see* **Chapter 9** in the companion volume, *Directed Enzyme Evolution: Screening and Selection Methods*).
6. Primers should be 30–50 bp, long depending on the number of mismatches included. Mismatches preferrably lie in the middle of the primer. Between the 3' end of the primer and any mismatch there should be at least 10 bp of matching nucleotides to get proper annealing. The 5' end is less critical.
7. The mutagenic primers can be designed as complements to one another. They can also be offset relative to each other to produce longer homologous ends. The length of homology should be at least 15 bp.
8. The *N*-terminal nonmutagenic primer is paired with the most *N*-terminal mutagenic antisense primer in one PCR, the corresponding sense primer is paired with the next antisense primer downstream in another reaction, and so on. The low concentration of template DNA avoids contamination of the transformation with parent sequence.
9. Yields of the gel extraction are usually higher with the concentrated sample than with the unpurified PCR reaction.
10. The relative amounts of the different PCR products may vary.

Acknowledgments

This work was supported by the US Office of Naval Research. We thank the Ministerio de Educacion y Cultura of Spain (MA) and Deutsche Forschungsgemeinschaft (TB) for fellowships.

References

1. Sung, P., Trujillo, K. M., and Van Komen, S. (2000) Recombination factors of *Saccharomyces cerevisiae. Mutat. Res. Fund. Mol. Mech. Mutagen.* **451,** 257–275.
2. Orrweaver, T. L., Szostak, J. W., and Rothstein, R. J. (1981) Yeast transformation - a model system for the study of recombination. *Proc. Natl. Acad. Sci. USA-Biological Sciences* **78,** 6354–6358.
3. Muhlrad, D., Hunter, R., and Parker, R. (1992) A rapid method for localized mutagenesis of yeast genes. *Yeast* **8,** 79–82.
4. Ma, H., Kunes, S., Schatz, P. J., and Botstein, D. (1987) Plasmid construction by homologous recombination in yeast. *Gene* **58,** 201–216.
5. Cherry, J. R., Lamsa, M. H., Schneider, P., et al. (1999) Directed evolution of a fungal peroxidase. *Nat. Biotechnol.* **17,** 379–384.

6. Morawski, B., Lin, Z., Cirino, P., Joo, H., Bandara, G., and Arnold, F. H. (2000) Functional expression of horseradish peroxidase in *Saccharomyces cerevisiae* and *Pichia pastoris. Protein Eng.* **13,** 377–384.
7. Morawski, B., Quan, S., and Arnold, F. H. (2001) Functional expression and stabilization of horseradish peroxidase by directed evolution in *Saccharomyces cerevisiae. Biotechnol. Bioeng.* **76,** 99–107.
8. Abecassis, V., Pompon, D., and Truan, G. (2000) High efficiency family shuffling based on multi-step PCR and in vivo DNA recombination in yeast: statistical and functional analysis of a combinatorial library between human cytochrome P450 1A1 and 1A2. *Nucl. Acids Res.* **28,** E88.
9. Pompon, D. and Nicolas, A. (1989) Protein engineering by cDNA recombination in yeasts: shuffling of mammalian cytochrome P-450 functions. *Gene* **83,** 15–24.
10. Crameri, A., Raillard, S. A., Bermudez, E., and Stemmer, W. P. (1998) DNA shuffling of a family of genes from diverse species accelerates directed evolution. *Nature* **391,** 288–291.
11. Zhao, H., Giver, L., Shao, Z., Affholter, J. A., and Arnold, F. H. (1998) Molecular evolution by staggered extension process (StEP) in vitro recombination. *Nat. Biotechnol.* **16,** 258–261.
12. Stemmer, W. P. C. (1994) DNA Shuffling by random fragmentation and reassembly - in vitro recombination for molecular evolution. *Proc. Natl. Acad. Sci. USA* **91,** 10,747–10,751.
13. Jones, D. H., Riley, A. N., and Winistorfer, S. C. (1994) Production of a vector to facilitate DNA mutagenesis and recombination. *Biotechniques* **16,** 694–701.
14. Bulter, T., Alcalde, M., Sieber, V., et al. (2003) Functional expression of a fungal laccase in *S. cerevisiae* by directed evolution. *Appl. Environ. Microb.*, in press.
15. Kirkpatrick, R. D., Parchaliuk, D. L., WooQIAEX II, Qiagends, R. A., and Gietz, R. D. (1998) Transformation of *Saccharomyces cerevisiae* by the lithium acetate/single-stranded carrier DNA/polyethylene glycol (LiAc/ss-DNA/PEG) protocol. *Technical Tips Online* (http://tto.trends.com).
16. Thompson, J. R., Register, E., Curotto, J., Kurtz, M., and Kelly, R. (1998) An improved protocol for the preparation of yeast cells for transformation by electroporation. *Yeast* **14,** 565–571.

4

Creating Random Mutagenesis Libraries by Megaprimer PCR of Whole Plasmid (MEGAWHOP)

Kentaro Miyazaki

1. Introduction

The conventional method for cloning a DNA fragment is to insert it into a vector and ligate it *(1)*. Although this method is commonly used, it is labor intensive because the ratio and concentrations of the DNA insert and the vector must be optimized. Even then, the resulting library is often plagued with unwanted plasmids that have no inserts or multiple inserts. These species have to be eradicated to avoid tedious screening, especially when producing random mutagenesis libraries.

MEGAWHOP is a novel cloning method of DNA fragments which was developed as a substitute for the problematic ligation approach *(2)*. In MEGAWHOP, the DNA fragment to be cloned is used as a megaprimer that replaces a homologous region in the template plasmid (*see* **Fig. 1**). After running whole plasmid PCR using the megaprimer, the resultant mixture is treated with a *dam*-methylated DNA specific restriction enzyme, *Dpn*I. The treatment enables specific elimination of the template plasmid, because the template plasmid, which can be propagated in most *Escherichia coli* strains, is dam-methylated. The *Dpn*I-treated mixture is then introduced into *E. coli* to yield a library. Libraries produced by the MEGAWHOP method are virtually free from contamination by species without any inserts or with multiple inserts. A MEGAWHOP protocol that is especially ideal for creating random mutagenesis libraries is described below. The protocol guarantees a good outcome with a 500–1000 bp megaprimer and a 3–7 kbp template plasmid.

From: *Methods in Molecular Biology, vol. 231: Directed Evolution Library Creation: Methods and Protocols*
Edited by: F. H. Arnold and G. Georgiou © Humana Press Inc., Totowa, NJ

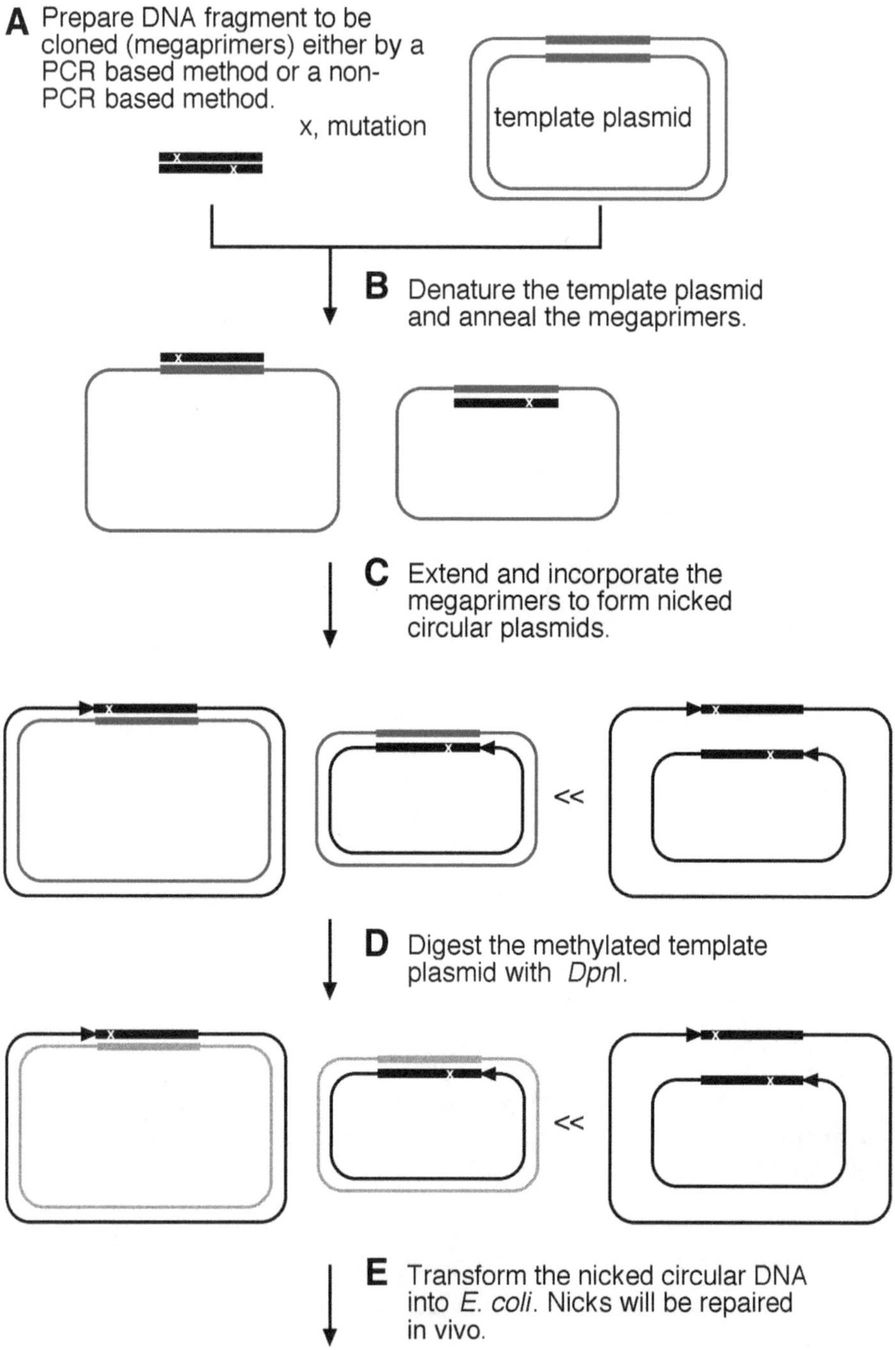
A Prepare DNA fragment to be cloned (megaprimers) either by a PCR based method or a non-PCR based method.
x, mutation
template plasmid
B Denature the template plasmid and anneal the megaprimers.
C Extend and incorporate the megaprimers to form nicked circular plasmids.
<<
D Digest the methylated template plasmid with *Dpn*I.
<<
E Transform the nicked circular DNA into *E. coli*. Nicks will be repaired in vivo.

2. Materials

1. Competent cells of *E. coli* (*dam*$^+$ strain, e.g., JM109, XL10).
2. Growth medium for *E. coli* (e.g., Luria Broth (LB), 2xYT).
3. Appropriate antibiotics (e.g., ampicillin, kanamycin, chloramphenicol, tetracycline).
4. DNA plasmid template of known concentration propagated in *E. coli dam*$^+$ strain and prepared using standard protocols ***(1)***.
5. *Dpn*I restriction enzyme (New England Biolabs; Beverly, MA).
6. 10X *PfuTurbo*® DNA polymerase buffer: 200 m*M* Tris-HCl, pH 8.8, 100 m*M* KCl, 100 m*M* $(NH_4)_2SO_4$, 20 m*M* $MgSO_4$, 1% Triton® X-100, 1 mg/mL bovine serum albumin (Stratagene; La Jolla, CA).
7. *PfuTurbo*® DNA polymerase (Stratagene).
8. dNTP set (Amersham; Piscataway, NJ).
9. Reagents and apparatus for agarose gel electrophoresis.
10. Qiaquick PCR purification kit (Qiagen, Hilden, Germany).
11. Thermal cycler (e.g., GeneAmp® 9700, Applied Biosystems; Foster City, CA).

3. Methods

The MEGAWHOP cloning method consists of four steps: (1) preparation of the DNA fragment to be cloned (megaprimer); (2) whole plasmid PCR using the megaprimer; (3) *Dpn*I-treatment of the whole plasmid PCR product; (4) and transformation, as described under **Subheadings 3.1–3.4.**

3.1. Preparation of DNA Fragment to be Cloned (Megaprimer)

1. Prepare a mutated gene fragment by random mutagenesis, using either a PCR-based procedure, e.g., error-prone PCR ***(3)***, DNA shuffling ***(4)***, or StEP recombination ***(5)*** (*see* **Note 1**) or a non-PCR-based procedure, e.g., chemical mutagenesis ***(1)***. Dissolve the DNA in water (or in Tris-EDTA (TE) buffer).
2. Determine the concentration of the DNA either by spectrophotometry (1 OD_{260} = 50 μg/mL) or by analytical agarose gel electrophoresis (use known concentration of DNA as standard).

3.2. Whole Plasmid PCR

1. Mix megaprimer and plasmid template: 0.5 μg megaprimer (*see* **Note 2**), 50 ng template plasmid (*see* **Note 3**), 0.2 m*M* of each dNTP, and 2.5 U of *PfuTurbo*®

Fig. 1. *(opposite)* Outline of the MEGAWHOP method. First, the mutated gene fragment is prepared by random mutagenesis. This can be done either by PCR-based (e.g., error-prone PCR, DNA shuffling, StEP recombination) or non-PCR-based procedures (e.g., chemical mutagenesis) **(A)** Fragments are then annealed to a template plasmid propagated in *E. coli dam*$^+$ strain **(B)** Next, whole plasmid PCR amplification is carried out to synthesize a nicked circular plasmid **(C)** The product is then treated with *Dpn*I to eliminate the template plasmid **(D)** The *Dpn*I-treated mixture is then introduced into *E. coli* and nicks are repaired in vivo **(E)**.

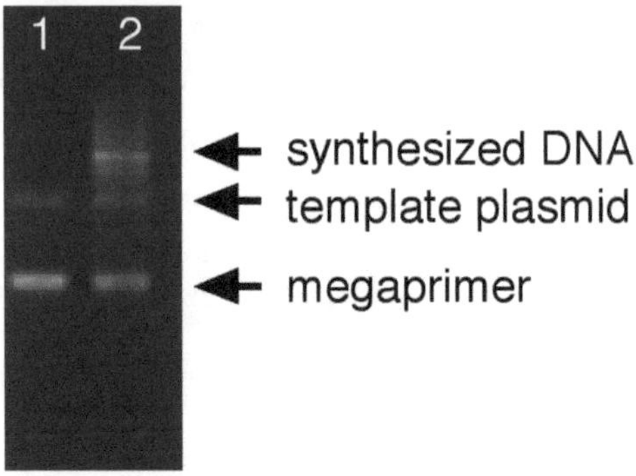

Fig. 2. Agarose gel electrophoresis of the MEGAWHOP reaction. 3 μL of the 50 μL whole plasmid PCR (lane 1, before; lane 2, after) were loaded onto a 1% agarose gel.

DNA polymerase (*see* **Note 4**) in 1x *PfuTurbo*® DNA polymerase buffer in a total volume of 50 μL.

2. Run whole plasmid PCR (*see* **Note 5**). Incubate the reaction mixture at 68°C for 5 min (this step is optional, *see* **Note 1**) and heat at 95°C for 1 min. Perform 18 cycles of incubation (*see* **Note 6**) at 95°C for 30 s, 55°C for 30 s, and 68°C for 6 min, using a GeneAmp® 9700 thermal cycler or equivalent apparatus. You can monitor the reaction by agarose gel electrophoresis as shown in **Fig. 2**.

3.3. DpnI Treatment

1. Following whole plasmid PCR, place the reaction on ice for 5 min to cool the reaction below 37°C.
2. Add 20 units of *Dpn*I (1 μL) directly into the whole plasmid PCR mixture (*see* **Notes 7** and **8**). Gently and thoroughly mix the reaction mixture by pipetting the solution up and down several times.
3. Incubate for 1–2 hr at 37°C.

3.4. Transformation

1. Use 1–2 μL of the *Dpn*I-treated mixture to transform 20–100 μL of competent *E. coli* cells (*see* **Note 9**) using standard protocols *(1)*. Grow the cells on an agar plate containing appropriate antibiotics.
2. Typically, the transform should yield ~500–1000 colonies when "standard" (i.e., 10^7/μg of pUC18) competent cells are used. Each plasmid (> 99.9%) should now be carrying a single insert.

4. Notes

1. If you prepare a fragment using *Taq* DNA polymerase (common in error-prone PCR), be sure to remove extra nucleotides at the 3'-end of the products *(7,8)* to avoid unintended mutation during whole plasmid PCR. This can be done by treat-

ing the fragment with a DNA polymerase having 3' to 5' exonuclease activity (e.g., T4 DNA polymerase, *Pfu* DNA polymerase). More conveniently, you can simply add a short incubation period (68°C for 5 min) prior to the whole plasmid PCR program because the reaction is usually performed with a high-fidelity polymerase having 3' to 5' exonuclease activity.

2. Increasing the concentration of megaprimer increases the yield of synthesized plasmid and the number of transformants. The yield reaches a plateau when 0.2 μg of megaprimer (~ 750 bp) and 50 ng of plasmid template (~ 3.5 kbp) are used. No significant increase in the yield is observed even if a larger amount of megaprimer is used.
3. Increasing the concentration of template plasmid increases the yield of synthesized plasmid and the number of transformants. The yield reaches a plateau when 25 ng of template plasmid (~ 3.5 kbp) and 0.5 μg of megaprimer (~ 750 bp) are used. No significant increase in the yield is observed even if a larger amount of template plasmid is used.
4. In order to minimize the incorporation of new mutations during the long-range PCR, it is essential to use a high fidelity DNA polymerase. We used *PfuTurbo*® DNA polymerase. You may also use other types of high fidelity DNA polymerase such as native *Pfu* DNA polymerase (Stratagene), Vent® DNA polymerase (New England Biolabs), or KOD DNA polymerase (Toyobo; Tokyo, Japan).
5. The extension time can be calculated as 2 min/kb, based on the length of the DNA template using *PfuTurbo*® DNA polymerase. If you use a different DNA polymerase, the time has to be optimized. Follow the instructions supplied with the polymerase you choose.
6. Increasing the number of cycles increases the yield of synthesized plasmid and the number of transformants. In our case, decreasing the number of cycles to 12 decreases the yield of transformants to approximately one-third. On the other hand, increasing the number of cycles to 24 increases the yield of transformants about twofold.
7. DpnI digests only methylated and hemi-methylated DNA *(6)*. Therefore, newly synthesized DNA using "standard" dNTP will not be digested by the *Dpn*I. On the other hand, template plasmid propagated in common *E. coli* is *dam*-methylated and hence will be digested with the restriction enzyme.
8. Most PCR buffers do not interfere with the *Dpn*I reaction.
9. The maximum yield of transformants is obtained when 0.5 μg of megaprimer (~ 750 bp) and 50 ng of template plasmid (~ 3.5 kbp) are used in 24 cycles of whole plasmid PCR. Under these conditions, more than 1×10^5 transformants are routinely obtained by using 1 μg of megaprimer and 100 μL of standard *E. coli* competent cells (~10^7/μg of pUC18). Virtually all (>99.9%) of the screened clones should contain a single insert.

Acknowledgments

I thank Misa Takenouchi for technical assistance. I also thank Professor Frances Arnold and her colleagues for helpful discussions.

References

1. Sambrook, J. and Russell, D. W. (2001) *Molecular Cloning: A Laboratory Manual, 3rd ed.* Cold Spring Harbor Laboratory Press, Cold Spring Harbor, NY.
2. Miyazaki, K. and Takenouchi, M. (2002) Creating random mutagenesis libraries using megaprimer PCR of whole plasmid. *Biotechniques* **33,**1033–1038.
3. Cadwell, R. C. and Joyce, G. F. (1992) Randomization of genes by PCR mutagenesis. *PCR Methods Appl.* **2,** 28–33.
4. Stemmer, W. P. (1994) Rapid evolution of a protein in vitro by DNA shuffling. *Nature* **370,** 389–391.
5. Zhao, H., Giver, L., Shao, Z., Affholter, J. A., and Arnold, F. H. (1998) Molecular evolution by staggered extension process (StEP) in vitro recombination. *Nat. Biotechnol.* **16,** 258–261.
6. McClelland, M. and Nelson, M. (1992) Effect of site-specific methylation on DNA modification methyltransferases and restriction endonucleases. *Nucleic Acids Res.* **20,** 2145–2157.
7. Clark, J. M. (1988) Novel non-templated nucleotide addition reactions catalyzed by procaryotic and eucaryotic DNA polymerases. *Nucleic Acids Res.* **16,** 9677–9686.
8. Hu, G. (1993) DNA polymerase-catalyzed addition of nontemplated extra nucleotides to the 3' end of a DNA fragment. *DNA Cell Biol.* **12,** 763–770.

5

Construction of Designed Protein Libraries Using Gene Assembly Mutagenesis

Paul H. Bessette, Marco A. Mena, Annalee W. Nguyen, and Patrick S. Daugherty

1. Introduction

Whole genes, plasmids, and even viral genomes can be assembled from relatively short, synthetic, overlapping oligodeoxyribonucleotides (oligos) by DNA polymerase extension *(1,2)*. Once the initial investment in a set of oligos spanning the length of a gene has been made, a mutation can be generated easily at any point in the gene merely by replacing two oligos and reassembling the gene. Swapping out the oligos encoding one or more amino acid residues (and their complementary overlapping oligos) with new degenerate oligos allows for libraries of mutants (*see* **Fig. 1**) *(3)*. By this method, multiple positions can be targeted in a single step with no limitations on their proximity, and every targeted position is guaranteed to be mutated in the product. Diversity is generated entirely in vitro, and the protocol, involving two rounds of PCR with no gel purifications, can be accomplished easily in less than one day, followed by cloning as one would for a traditional PCR product.

While many diversity generating methods exist for creating DNA-level libraries of variants, each has advantages and disadvantages for particular applications. In addition to the simplicity of the method herein and its ability to target multiple sites, assembly PCR benefits from the complete absence of template contamination in the final product, since none is ever present. PCR template in traditional cassette mutagenesis and overlap exchange PCR methods can be extremely difficult to remove, and, thus, can lead to background artifacts and false positives. This issue is especially important when searching for rare library members in large libraries, where even low-level contamination with unmutated sequences can be a problem.

From: *Methods in Molecular Biology, vol. 231: Directed Evolution Library Creation: Methods and Protocols*
Edited by: F. H. Arnold and G. Georgiou © Humana Press Inc., Totowa, NJ

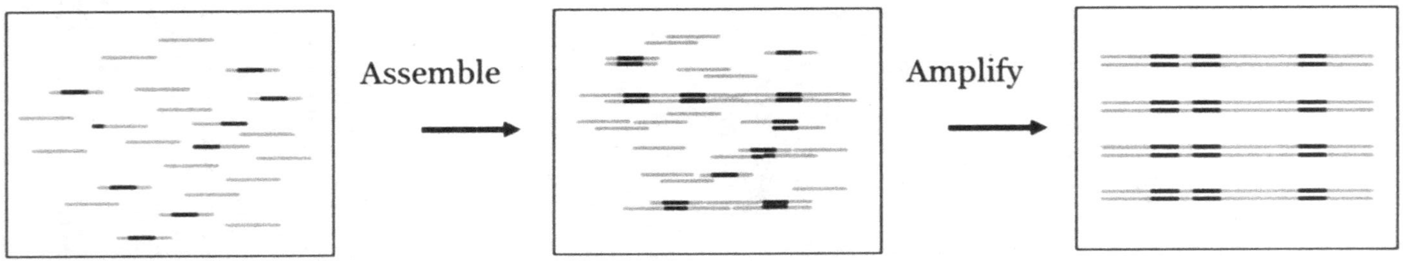

Fig. 1. Schematic of gene assembly mutagenesis. A population of gene variants is assembled from short oligonucleotides encoding both strands of the gene and containing degenerate bases at the targeted positions (shown in bold). Following assembly PCR, the full length gene variants are amplified using the outside primers.

The assembly mutagenesis method has a technical limitation of introducing non-targeted mutations at an elevated rate relative to routine PCR. While this fortuitous diversity may be desirable in some applications, it may be necessary to increase the library size to ensure complete representation of *all possible* intended sequences. As many as 75% of clones may contain errors (PHB unpublished), indicating that a library should be roughly fourfold larger than otherwise needed. The gene assembly protocol originally used by Stemmer et al. ***(1)*** employing a mixture of *Taq* and *Pfu* polymerases results in 7–9 point mutations per kilobase of DNA (***4***, PHB unpublished). Though the exclusive use of "proof-reading" polymerases that contain 3' to 5' exonuclease activity results in an improvement over the *Taq/Pfu* mixture to 0.6–3.5 errors per kilobase (***4,5***, AWN, PHB unpublished), the error frequency is still much higher than expected for amplification in routine PCR methods using these enzymes. The quality of oligos, number and length of oligos used, and number of cycles used can collectively influence the fidelity of gene synthesis, yet the assembly process itself may be inherently error-prone.

Frame-shifted mutants, consisting almost exclusively of single base deletions, represent a second class of errors in the assembled PCR products, arising at a frequency of 0.4–1.3 per kilobase (AWN, PHB unpublished). As these presumably arise from inefficiency of coupling in the solid-phase synthesis of oligonucleotides, PAGE purification of individual oligos would likely alleviate the frequency of deletion mutants. This option may not be economically feasible, however, and might not be necessary, since frame-shifted mutants are unlikely to give rise to false positives. On the other hand, to maximize the functional library size, in some cases it may be possible to positively select for in-frame translation by fusion to a reporter gene (e.g., chloramphenicol acetyl transferase) ***(6)***.

The protocol typically involves assembling the gene of interest from a series of 40-mer oligos encoding both strands, with the oligo junctions on top and bottom strands offset by 20 bases. One or more pairs of complementary oligos are heterogeneous at the positions targeted for mutagenesis. The oligos are hybridized and extended in multiple cycles, with growing strands eventually overlapping and comprising the full gene length. The products are diluted into a second PCR with only the outermost primers present in high concentration, such that the full length gene is preferentially amplified. Inclusion of *Sfi*I restriction sites on either side of the gene and 50–70 bases of flanking sequence in the assembly allows for efficient restriction digestion of the final product. The digested products can then be ligated into a regulated expression vector that contains two compatible *Sfi*I sites. The use of *Sfi*I allows for directional cloning of inserts using a single restriction enzyme, and the non-palindromic overhangs created disallow vector dimers, improving cloning efficiency.

2. Materials

1. 40-mer oligodeoxyribonucleotides (oligos).
2. 10 m*M* Tris-HCl, pH 8.0.
3. KOD Hot Start Polymerase (Novagen, Madison, WI).
4. dNTPs : 2.0 m*M* each.
5. 25 m*M* $MgSO_4$.
6. KOD polymerase 10× buffer (Novagen, Madison, WI).
7. Thermal cycler.
8. Agarose gel electrophoresis equipment and supplies.
9. QIAquick PCR cleanup kit (QIAgen, Valencia, CA).
10. *Sfi*I restriction enzyme (New England Biolabs, Beverly, MA).
11. 10X NEBuffer 2: 500 m*M* NaCl, 100 m*M* Tris-HCl, pH 7.9, 100 m*M* $MgCl_2$, 10 m*M* dithiothreitol (DTT).
12. Bovine serum albumin (BSA) (1 μg/μL).
13. Regulated expression vector.
14. T4 DNA Ligase (1 U/μL) (Invitrogen, Carlsbad, CA).
15. 5X T4 DNA Ligase Buffer: 250 m*M* Tris-HCl, pH 7.6, *50* m*M* $MgCl_2$, 5 m*M* ATP, 5 m*M* DTT, 25% (w/v) polyethylene glycol-8000.
16. V Series nitrocellulose filter discs, 0.025 μm pore size (Millipore).
17. Electroporation apparatus and 0.2-cm gap cuvets.
18. Electrocompetent cells (>10^{10} colony-forming units [CFU]/μg).
19. SOC medium: 0.5% yeast extract, 2% tryptone, 10 m*M* NaCl, 2.5 m*M* KCl, 10 m*M* $MgCl_2$, 10 m*M* $MgSO_4$, 20 m*M* glucose.

3. Methods

3.1. Oligo Design

Design oligonucleotide primers 40 bases in length and encoding both the plus and minus strands of the entire gene, such that complementary oligos overlap by 20 bases. In place of the codons to be mutated, insert degenerate bases in the oligos on both complementary strands (*see* **Fig. 2B**, **Notes 1–3**). Include restriction sites or universal adapters in the flanking sequence for downstream cloning. Synthesize oligos at 10 or 50 nmol scale, desalted without further purification. Phosphorylation at the 5' ends is not necessary, as the assembly process does not involve ligation. Oligos are stored at –20°C as 100 μ*M* stock solutions in 10 m*M* Tris-HCl, pH 8.0.

3.2. Assembly PCR

1. Dilute oligos individually to 10 μ*M* in 10 m*M* Tris-HCl, pH 8.0.
2. Pool an equal volume of each oligo in a new tube.
3. Add the following to a 0.2 μL PCR tube: 4 μL oligo mixture, 2 μL 10X KOD polymerase buffer, 2 μL dNTPs (2.0 m*M* each stock solution), 0.8 μL $MgSO_4$ (25 m*M* stock solution), 10.2 μL PCR grade H_2O, 1 μL KOD Hot Start Polymerase (1 U/μL) (*see* **Note 4**).

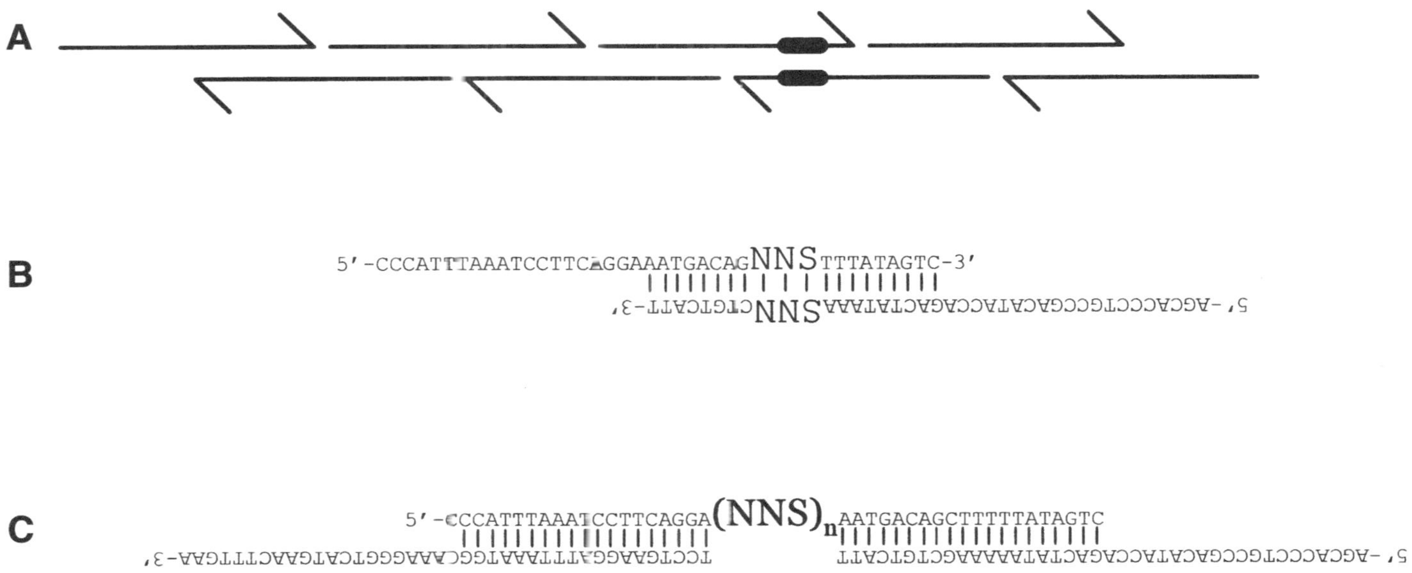

Fig. 2. **(A)** Oligos are designed to be complementary, with the exception of the 5' end of each strand, which includes a 20 bp overhang. Degenerate bases (bold lines) are encoded in both strands. **(B)** Close-up of a pair of mutagenic oligos using the NNS randomization scheme. **(C)** For linker length mutagenesis, a population of oligos can be used at the site of inserted codons, with the variable length region not complemented.

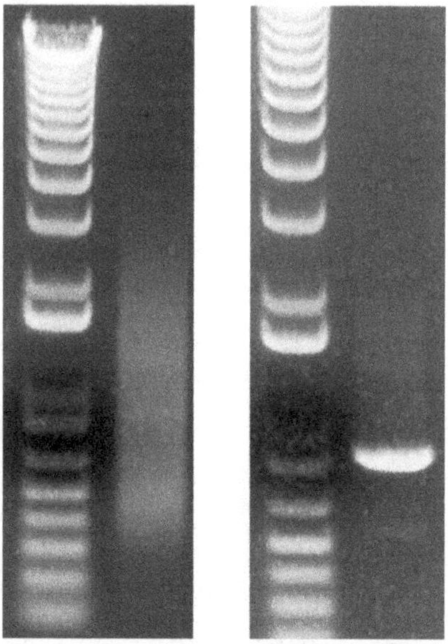

Fig. 3. (**A**) Following the assembly reaction, the products as visualized by agarose gel electrophoresis are a diffuse distribution spanning the expected size. Lane 1: Molecular weight markers; Lane 2: Assembly of a 900 bp gene from 44 oligos of 40 bases each. (**B**) Amplification of the assembly reaction using the outermost primers results in a single product of the expected size. Lane 1: Molecular weight markers; Lane 2: Amplification product of the assembly in **A**.

4. Mix the contents and place the tube in a thermal cycler with a heated lid.
5. Heat at 94°C for 2 min; 25 cycles of 94°C for 30 s, 52°C for 30 s, 68°C for 30 s; followed by a final step of 68°C for 10 min.
6. Check the assembly product by running 4 μL on an agarose gel stained with ethidium bromide. Expect to see a smear of products with a size distribution spanning the expected length of the gene (*see* **Fig. 3A**, **Notes 5** and **6**).

3.3. Product Amplification

The primers used for amplification are the outermost oligos from the 5' end of each strand.

1. Add the following to a 0.2 μL PCR tube: 1 μL Assembly product from above, 5 μL 10× KOD polymerase buffer, 5 μL dNTPs (2.0 m*M* each stock solution), 2 μL $MgSO_4$ (25 m*M* stock solution), 5 μL Forward primer (10 μ*M* stock), 5 μL Reverse primer (10 μ*M* stock), 26 μL PCR grade H_2O, 1 μL KOD Hot Start Polymerase (1 U/μL).
2. Mix the contents and place the tube in a thermal cycler with a heated lid.

3. Heat at 94°C for 2 min; 23 cycles of 94°C for 30 s, 55°C for 30 s, 68°C for 1 min; followed by a final step of 68°C for 10 min.
4. Verify the final product by running 2 µL on an agarose gel stained with ethidium bromide (*see* **Fig. 3B**).
5. Purify the PCR product using the QIAquick PCR cleanup kit following the manufacturer's instructions, and elute into 40 µL of H_2O.

3.4. Cloning

1. Prepare the PCR product restriction digestion as follows: 40 µL PCR product in H_2O, 5.5 µL 10× NEBuffer 2, 5.5 µL 10_ BSA (1 µg/µL), 4 µL *Sfi*I (20 U/µL).
2. Incubate the reaction at 50°C for 3 h.
3. Run the entire reaction on an agarose gel; cut out the band of the correct size, and recover the DNA using the QIAquick gel extraction kit following the manufacturer's instructions.
4. Quantitate the concentration of the PCR product and ligate into the expression vector, which has been previously digested with *Sfi*I and dephosphorylated. Use a ratio of 30 fmol vector ends to 90 fmol PCR product ends, where

$$fmol.ends/ng = (2 \times 10^6)/[660 \times (\#bp)]$$

while maintaining the total mass of DNA less than 100 ng.
5. The ligation reaction volume should be 20 µL total with vector, insert, 4 µL of 5X T4 DNA ligase buffer (containing polyethylene glycol), 1 µL (1 U) of T4 DNA ligase, and sterile H_2O. Incubate the ligation reaction for 1 h at room temperature (23°C) or overnight at 14°C (*see* **Note 7**).
6. Deactivate the ligase by incubating the ligation at 70°C for 10 min.
7. Desalt the ligation reaction by filling a sterile disposable Petri dish with sterile water and floating a 0.025 µm nitrocellulose filter on the water with the shiny side up. Deposit the ligation reaction on the filter and allow it to sit undisturbed for 1 h at room temperature. This will result in an increase in volume to approximately 30 µL.
8. Recover the ligation from the filter with a micropipette and chill it in a microcentrifuge tube on ice. Chill a 0.2 cm electroporation cuvette on ice. Thaw electrocompetent cells (>10^{10} cfu/µg) on ice.
9. As soon as electrocompetent cells are thawed, add 200 µL cells to the tube containing the ligated DNA; mix by pipetting once, and transfer all to the cuvet. Pulse the cuvet at 2.5 kV, 100 Ω, 50 µF. Immediately wash the cuvet three times with 1 mL of SOC medium (pre-warmed to 37°C), and transfer the cells to a 125-mL culture flask. Add an additional 7 mL of SOC and incubate the flask with shaking at 37°C for 1 h before adding antibiotic for selection.
10. Dilute an aliquot of cells, and plate the cells on selective medium for overnight growth in order to assay the library size.

4. Notes

1. Assembling the entire gene from synthetic oligos allows the researcher great flexibility in optimizing the codon usage for the desired expression host, as well as

adding or removing restriction sites and genetic markers by silent mutagenesis or in-frame insertions/deletions.

2. Hoover and Lubkowski have developed useful software to automate the design of oligos for whole gene assembly PCR *(5)*. The freely available DNAWorks program performs back translation according to user defined codon preference for the expression system of choice, and additionally defines primer junctions and optimizes the overlapping regions for homogenous melting temperatures and lack of hairpin formation.
3. Mispriming can inhibit successful assembly. Ideally oligos should be scrutinized and silent mutagenesis employed to circumvent the tendency for incorrect dimer formation. The use of an enzyme mix containing hot start antibody or manual hot start (adding polymerase after the initial denaturation step) is recommended. Synthesis of long genes increases the chances for mispriming, but can be accomplished alternatively by splitting the assembly PCR into two or more reactions each containing only the complementary oligos for a portion of the gene, such that the products will span the entire gene and overlap by 20 or more bases. Then, in the amplification reaction add 1 μL from each assembly along with the two outside primers to reconstitute the gene by overlap exchange.
4. Other proof-reading polymerases may be employed, though lack of terminal transferase activity is thought to be important for higher fidelity in this method. Therefore, enzyme mixtures containing *Taq* polymerase or other enzymes that do not leave blunt-ended PCR products are not recommended.
5. The values given for number of cycles, annealing temperature, magnesium concentration, and extension times are guidelines that will vary based on the application. In particular, long genes or the use of polymerase with lower processivity will require longer extension times and possibly more cycles. For especially long assemblies, Stemmer and others have employed two rounds of assembly, where the first round reaction is diluted into fresh PCR mix for additional cycles, up to 100 or more total cycles *(1,7)*.
6. Although assembly of smaller genes and fragments (<300 bp) may give rise to a single product in the initial assembly reaction, the normal situation with larger sized assemblies presents a smear of products distributed from lower to much larger molecular weight than the expected size. Though the composition of the higher molecular weight products has not been examined, the second round reaction with only the outside primers usually resolves a single product.
7. Larger libraries can be created by scaling up the ligation reaction, maintaining the same ratio of DNA mass to reaction volume, and at least 5 U/mL final concentration of ligase. After heat inactivation of the ligase, the ligation reaction is then ethanol precipitated and resuspended in 20–40 μL of water before dialysis on the nitrocellulose membrane. More than one transformation may be needed.
8. To mitigate biases resulting from the stochastic nature of degenerate primer annealing in the early rounds of assembly, multiple reactions (10 or more) can be assembled in parallel and pooled before restriction enzyme digestion.

9. In addition to directed point mutagenesis, the assembly PCR method allows for variability of encoded polypeptide length, such as may be desired in library screening of flexible linkers or antibody CDRs. Oligos can be designed as in **Fig. 2C**, such that multiple oligos encode the plus strand at the position of inserted codons, containing 20 bases on each end overlapping the complementary oligos, and an additional 3, 6, 9, etc. bases that are not complemented by the two overlapping oligos.

References

1. Stemmer, W. P., Crameri, A., Ha, K. D., Brennan, T. M., and Heyneker, H. L. (1995) Single-step assembly of a gene and entire plasmid from large numbers of oligodeoxyribonucleotides. *Gene* **164,** 49–53.
2. Cello, J., Paul, A. V., and Wimmer, E. (2002) Chemical Synthesis of Poliovirus cDNA: Generation of Infectious Virus in the Absence of Natural Template. *Science* **297,** 1016–1018.
3. Silverman, J. A., Balakrishnan, R., and Harbury, P. B. (2001) Reverse engineering the (beta/alpha)8 barrel fold. *Proc. Natl. Acad. Sci. USA* **98,** 3092–3097.
4. Withers-Martinez, C., Carpenter, E. P., Hackett, F., et al. (1999) PCR-based gene synthesis as an efficient approach for expression of the A+T-rich malaria genome. *Protein Eng.* **12,** 1113–1120.
5. Hoover, D. M. and Lubkowski, J. (2002) DNAWorks: an automated method for designing oligonucleotides for PCR-based gene synthesis. *Nucleic Acids Res.* **30,** e43.
6. Sieber, V., Martinez, C. A., and Arnold, F. H. (2001). Libraries of hybrid proteins from distantly related sequences. *Nat. Biotechnol.* **19,** 456–460.
7. Baedeker, M. and Schulz, G. E. (1999) Overexpression of a designed 2.2 kb gene of eukaryotic phenylalanine ammonia-lyase in *Escherichia coli. FEBS Lett.* **457,** 57–60.

6

Production of Randomly Mutated Plasmid Libraries Using Mutator Strains

Annalee W. Nguyen and Patrick S. Daugherty

1. Introduction

A variety of methods have been developed for random mutagenesis of genes and whole plasmids to generate genetic diversity for directed evolution experiments *(1)*. In particular, bacterial strains exhibiting unusually high rates of spontaneous DNA mutagenesis, or mutator strains, can be used to generate large, diverse plasmid libraries. Propagation of plasmids within relatively stable mutator strains has been shown to provide a rich spectrum of single base substitutions, insertions, and deletions throughout the entire plasmid *(2)*. A significant advantage of using mutator strains to create random libraries is the accessibility of the procedure. Specialized equipment or cloning techniques are not required. Researchers can potentially create a large random library in two or three days, with minimal effort or prior experience in recombinant DNA techniques.

Mutator strains are typically characterized by their genetic deficiencies in DNA proofreading and editing machinery. The most commonly occurring deficiencies involve mutations in the *mutD*, *mutS*, and *mutT* genes. Specifically, *mutD* mutations can hinder the 3'-5' exonuclease activity of DNA polymerase III, thereby preventing repair of incorrectly incorporated bases *(3)*. The resulting mutations are predominantly transitions (85%), but also include transversions (10%), and frameshifts (5%) *(2,4)*. In minimal media, *mutD* strains exhibit a mutation rate 10 to 100 times that of wild-type, while cultures in rich media have up to 10^3 to 10^5-fold increase in mutation rate over wild-type *(2)*. The *mutS* mutation disables DNA mismatch repair, resulting in transitions and transversions *(5)*. Finally, *mutT* mutations prevent the degradation of 8-oxodGTP in mismatches involving A:G *(6)*, resulting in mainly AT-CG transversions *(2)*. While many bacterial strains with *mutD* mutations have been

From: *Methods in Molecular Biology, vol. 231: Directed Evolution Library Creation: Methods and Protocols*
Edited by: F. H. Arnold and G. Georgiou © Humana Press Inc., Totowa, NJ

used for mutagenesis (e.g., GM4708 *[7]*, CC954 *[8]*, mutD5-FIT *[9]*), the widely used and commercially available mutator strain, XL1-RED (Stratagene; La Jolla, CA) includes *mutD*, *mutS*, and *mutT* mutations. The combination of these three alleles elevates mutagenesis rates to on average 0.5 mutations per kb after 30 generations of growth in XL1-RED ***(10)***.

While mutator strains provide significant advantages in some applications, they possess a few notable drawbacks. First, the mutator phenotype is intrinsically unstable and must be monitored. The mutations that cause higher frequency mutagenesis are also responsible for slow growth rates of the mutator strains (e.g., doubling time of about 90 min ***[10]***). Thus, clones with reduced mutator phenotypes will quickly overgrow a culture, thereby preventing further mutagenesis. Second, multiple days of growth are required for more than a couple of mutations per gene. High-frequency mutagenesis may be required for faster evolutionary improvement ***(11)***, in which case, PCR-based mutagenesis is less time consuming. Lastly, the inability to target mutations within a single gene may present problems in evolutionary selections where changes in promoter sequence and the plasmid origin may result in increases in protein production, rather than improvement of the activity of interest ***(10)***.

The following protocol outlines the basic procedure to create a library using a bacterial mutator strain.

2. Materials

1. *E. coli* mutator strain (frozen stock or heat shock competent).
2. M9 minimal medium: 1X M9 salts, pH 7.4, 2 m*M* $MgSO_4$, 100 μ*M* $CaCl_2$, 0.4% D-(+)-glucose, 5 μg/mL thiamine-HCl, 1% casamino acids.
3. M9 minimal medium plates: M9 minimal medium plus 15 g/L agar.
4. Plasmid DNA including the gene of interest (*see* **Note 4**).
5. Antibiotic.
6. Chemical inducer.
7. β-mercaptoethanol.
8. SOC media.
9. Luria Bertani (LB) Broth.
10. Sterile plastic loop or spatula.
11. Mini-prep kit (Qiagen; Valencia, CA), or equivalent.
12. 10 m*M* Tris-HCl, pH 8.5.
13. V Series nitrocellulose filter discs, 0.025 μm pore size (Millipore; Bedford, MA).
14. 0.1 cm gap electroporation cuvet and electroporation apparatus.
15. Electrocompetent cells (non-mutator).
16. Sequencing primer.

3. Methods

The following methods describe 1) the preparation of competent mutator cells, 2) the transformation of the mutator strain with a plasmid, 3) the growth of the mutator strain resulting in mutagenesis, 4) the transfer of the mutated plasmid to a stable strain, and 5) the initial analysis of the library.

3.1. Preparing Competent Cells

1. Streak mutator strain to M9 minimal medium plates and grow at 37°C to obtain single colonies (*see* **Note 1**).
2. After 36 h, colonies will be rough-edged and various sizes. Choose approximately 20 colonies at random for inoculation of a single 400 mL M9 minimal medium culture.
3. Grow the 400 mL mutator strain culture to an OD_{600} of 0.35–0.38.
4. Make heat shock competent cells according to the *Current Protocols in Molecular Biology* procedure *(12)*.

3.2. Transformation of Mutator Strains (see Note 2)

1. Thaw 50 µL heat shock cells per transformation on ice.
2. Pre-heat SOC to 42°C.
3. For each transformation, aliquot 50 µL of cells to a 1.5 mL eppendorf tube and place on ice (*see* **Note 3**).
4. Add β-mercaptoethanol to the cells to a final concentration of 25 m*M*.
5. Gently swirl the tube to mix and incubate cells on ice for 10 min.
6. Add 10–50 ng of plasmid DNA (*see* **Note 4**) in less than 5 µL to the cells and incubate the mixture on ice for approximately 30 min.
7. Transfer the tubes to a 42°C water bath for 45 s.
8. Incubate the tubes on ice for 2 min.
9. Add 1 mL of SOC, pre-warmed to 42°C, to each tube and transfer the entire contents to a culture tube. Shake culture (250 rpm) at 37°C for 1 h.
10. Plate 200 µL aliquots of each transformation on LB plates with antibiotic and repressor appropriate to the plasmid.
11. Incubate the plates at 37°C for 30 h.

3.3. Propagation of the Mutator Strains

1. More than 100 colonies should be present on the plate containing 200 µL of transformation culture. Add 1–2 mL of LB to the 200 µL plate and scrape the colonies from the agar with a sterile plastic loop or spatula. Resuspend the cells in LB on the plate by pipetting up and down. Transfer the cell suspension to 10 mL of LB.
2. Measure the OD_{600} of the cell mixture and dilute the cells in 200 mL LB, with appropriate antibiotics, to a calculated OD_{600} of 0.0005 (*see* **Note 5**).
3. Monitor the OD_{600} of the culture to determine the number of generations the cells have been grown (*see* **Note 6**). This is determined using the equation:

$$N = N_0 \times 2^n$$

where N is the final number of cells, N_0 is the initial number of cells and n is the number of generations.

4. The culture should be rediluted while it is still in log phase to minimize overgrowth of non-mutator cells. An appropriate point for dilution is at an OD_{600} between 2.0 and 2.5 (approx 12 generations). Redilute the culture to a calculated OD_{600} of 0.0005 in 200 mL of LB with antibiotic and repressor.
5. Repeat dilution and growth of the cells until the desired number of generations is reached (fewer than 35 generations is recommended). To obtain a high yield of plasmid DNA, adjust the culture volume and dilution factor to ensure that the culture will be near saturation when the cells are harvested.

3.4. Transfer of the Library to a Stable Strain

1. After growth to the desired number of generations, prepare a plasmid DNA miniprep using 3 mL of culture, and elute the plasmid into 50 µl of 10 m*M* Tris-HCl buffer, pH 8.5 (*see* **Note 7**).
2. Place a nitrocellulose filter "shiny side up" on sterile water in a petri dish. Place 5 µL of the isolated supercoiled plasmid on the filter and leave undisturbed for 1–2 h to desalt the DNA.
3. The volume of the droplet containing plasmid will increase over the desalting period. Remove 5 µL of DNA from the filter and place in a 1.5 mL eppendorf tube on ice.
4. Chill an electroporation cuvet (0.1 cm gap) on ice.
5. Pre-heat SOC to 37°C.
6. Thaw 35 µL of electrocompetent cells on ice and add to the tube containing the desalted DNA.
7. Transfer the cell/DNA mixture to the chilled electroporation cuvet and remove any bubbles by gently tapping the bottom of the cuvette on the counter about 5 times. Leave the cuvet on ice.
8. Set the electroporation apparatus to 100 Ω, 50 µF, and 1.8 kV. Wipe off any water on the sides of the cuvet and place it in the electroporation chamber. Pulse the cells. The time constant should be near 5.0 ms.
9. Immediately after electroporation, rinse the cuvet twice with 0.5 mL SOC, prewarmed to 37°C. After each rinse, transfer the contents of the electroporation cuvet to a culture tube.
10. Incubate the transformed cells at 37°C with 250 rpm shaking for 1 h.

3.5. Analysis of Library

3.5.1. Size

1. Once the transformation has recovered for 1 h, make a 1:1000 dilution in 1 mL of SOC. Plate 100 µL and 10 µL aliquots of the 1:1000 dilution on LB plates with antibiotic and repressor. Add 100 µL of SOC to the 10 µL aliquot before spreading to ensure even dispersal of the smaller culture volume.
2. Grow overnight at 37°C.
3. Count the number of colonies on each plate and calculate the total library size represented in the transformation.

3.5.2. Diversity (This step is optional, but recommended)

1. Depending on the expected mutational load, individually culture 10–30 clones (isolated from plates in **Subheading 3.5.1.**) in 5 mL of LB with antibiotic and repressor.
2. Mini-prep each clone and sequence with an appropriate sequencing primer.
3. Analyze sequencing data to obtain the mutation rate (mutations/gene, % mutated).

3.5.3. Screening/Selection

3.5.3.1. Plates

1. After 1 h of recovery, the transformed library may be directly plated on appropriate growth or selection media (with inducer) and incubated as required.

3.5.3.2. Liquid

1. Add the remaining transformation mixture to 10 mL of LB with antibiotic (non-inducing conditions) for overnight growth.
2. Once the culture has reached saturation, subculture 1:100 in appropriate media with antibiotic and grow for 2–3 h, then induce as appropriate.

4. Notes

1. The *mutD* mutation is suppressed when grown on minimal media (owing to the absence of thymidine *[2]*). Any cell culture should be carried out on minimal media prior to introduction of the target plasmid to minimize the probability of loss of the mutator genotype.
2. The transformation protocol is adapted from literature provided by Stratagene with the purchase of XL1-Red cells.
3. Heat shock protocols often recommend the use of 100 μL cells and Falcon 2059 polypropylene tubes, but the authors have obtained high-efficiency transformations using 50 μL cells and typical 1.5 mL polypropylene eppendorf tubes.
4. The *mutD* mutation affects only DNA polymerase III, and the rate of mutagenesis is significantly lower when colE1-type origin plasmids are used. ColE1 origins of replication rely on DNA polymerase I for replication and should not be used for efficient mutagenesis of the target gene *(8)*.
5. The volume and dilution factor may be adjusted for convenience, but maintenance of a high level of library diversity is important. Each dilution will reduce the diversity of the library by selecting only a small portion of the entire population for further propagation. The guidelines described here involve further growth of approximately 10^8 cells per dilution.
6. Appropriate dilutions must be made to keep the OD readings in the range that correlates linearly with concentration. Normally this range is Abs 0.01 to 1.
7. If a higher mutational frequency is desired, the mini-prepped DNA can be retransformed to the mutator strain by repeating **Subheading 3.2–3.4., step 1** as required.

References

1. Brakmann, S. (2001) Discovery of superior enzymes by directed molecular evolution. *Chembiochem* **2,** 865–871.
2. Cox, E. C. (1976) Bacterial mutator genes and the control of spontaneous mutation. *Annu. Rev. Genet.* **10,** 135–156.
3. Miller, J. H. (1998) Mutators in *Escherichia coli. Mutat. Res.* **409,** 99–106.
4. Schaaper, R. M. (1988) Mechanisms of mutagenesis in the *Escherichia coli* mutator *mutD*5: role of DNA mismatch repair. *Proc. Natl. Acad. Sci. USA* **85,** 8126–8130.
5. Hsieh, P. (2001) Molecular mechanisms of DNA mismatch repair. *Mutat. Res.* **486,** 71–87.
6. Fowler, R. G. and Schaaper R. M. (1997) The role of the *mutT* gene of *Escherichia coli* in maintaining replication fidelity. *FEMS Microbiol. Rev.* **21,** 43–54.
7. Palmer, B. R. and Marinus M. G. (1991) DNA methylation alters the pattern of spontaneous mutation in *Escherichia coli* cells (*mutD*) defective in DNA polymerase III proofreading. *Mutat. Res.* **264,** 15–23.
8. Stefan, A., et al. (2001) Directed evolution of beta-galactosidase from *Escherichia coli* by mutator strains defective in the 3'→5' exonuclease activity of DNA polymerase III. *FEBS Lett.* **493,** 139–143.
9. Coia, G., et al. (1997) Use of mutator cells as a means for increasing production levels of a recombinant antibody directed against Hepatitis B. *Gene* **201,** 203–209.
10. Greener, A., Callahan, M., and Jerpseth, B. (1997) An efficient random mutagenesis technique using an *E. coli* mutator strain. *Mol. Biotechnol.* **7,** 189–195.
11. Daugherty, P.S., et al. (2000) Quantitative analysis of the effect of the mutation frequency on the affinity maturation of single chain Fv antibodies. *Proc. Natl. Acad. Sci. USA* **97,** 2029–2034.
12. Seidman, C.E., Struhl, K., Sheen, J., and Jessen, T. (1997) Introduction of Plasmid DNA into Cells, in *Current Protocols in Molecular Biology* (Ausubel, F.M., Brent, R., Kingston, R. E., et al., eds.), John Wiley & Sons Inc. Hoboken, NJ, pp. 1.8.1–1.8.2.

7

Evolution of Microorganisms Using Mutator Plasmids

Olga Selifonova and Volker Schellenberger

1. Introduction

The directed evolution of microorganisms requires the random introduction of mutations and subsequent selection of improved variants. Traditionally, mutagenesis has been performed using DNA-modifying chemicals or UV radiation. An alternative is the use of so-called mutator strains, which carry defects in one or several DNA repair genes. Such mutator strains have been used for the directed evolution of individual genes on plasmids or phagemids *(1,2)*. It is well documented that mutator strains can evolve rapidly to produce offsprings with novel phenotypes. However, their application for research or as industrial production hosts has been prevented by their very limited genetic stability.

Mutator plasmids allow one to temporarily convert a normal bacterial strain into a mutator strain by supplying the mutator gene on the plasmid *(3)*. This increases the mutation rate of a bacterial strain by about 20–4000 fold. In this mutator state, the strains can be rapidly evolved by culturing them under the appropriate selection pressure. The mutator plasmid carries a temperature-sensitive origin of replication which facilitates its rapid removal from the evolved strain yielding a stable strain with novel phenotype. This process of mutator based strain evolution is outlined in **Fig. 1**.

The mutator plasmid encodes a non-functional mutant of the *mutD* (*dnaQ*) gene. This gene encodes the ε-subunit of DNA polymerase III responsible for proofreading. The *mutD5* allele carries two amino acid mutations, and leads to the production of a protein that can effectively bind to DNA polymerase III. As a result, the *mutD5* protein can displace the functional allele of the same protein that is produced from the chromosomal allele of *mutD* *(4)*. This dominant

From: *Methods in Molecular Biology, vol. 231: Directed Evolution Library Creation: Methods and Protocols*
Edited by: F. H. Arnold and G. Georgiou © Humana Press Inc., Totowa, NJ

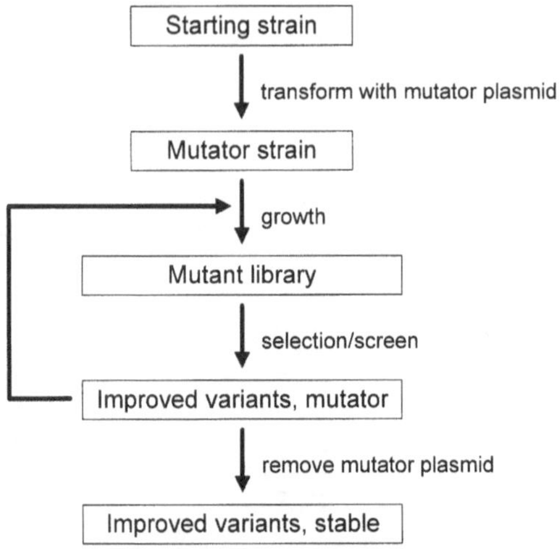

Fig. 1. Strain evolution using a mutator plasmid. The starting strain is transformed with the mutator plasmid. It is grown to introduce mutations and resulting mutant population is subjected to selection or screening to enrich improved variants. These two steps can be combined as mutagenesis will occur during enrichment. The process of mutagenesis and enrichment can be repeated multiple times until the desired phenotype is obtained. Subsequently, the mutator plasmid is removed to stabilize the new phenotype.

phenotype of the mutD5 allele allows one to convert a strain into a mutator strain without the need to manipulate its chromosome.

The actual process of mutagenesis and selection or screening is dependent on the desired trait to be evolved. Selection can be performed by serial transfer in liquid culture, continuous culture in a fermenter, or by serial transfer on solid medium. Variants can also be enriched using a fluorescence activated cell sorter (FACS). In the subsequent example, we describe the evolution of a DMF-resistant variant of *E. coli* by screening on agar plates.

2. Materials

1. MutD5 mutator plasmid.
2. *E. coli* CSH116 (as a source of *mutD5* gene).
3. Competent cells (*E. coli* MM294, *E. blattae* 33429, *see* example).
4. Oligonucleotide primers.
5. *Pfu Turbo* DNA polymerase (Stratagene, La Jolla, CA).
6. 10X cloned *Pfu* buffer: 200 m*M* Tris-HCl, pH 8.8, 20 m*M* $MgSO_4$, 100 m*M* KCl, 100 m*M* $(NH_4)_2SO_4$, 1% Triton X-100, 1mg/mL nuclease-free Bovine Serum Albumin (BSA).
7. T4 DNA polymerase (New England Biolabs, Beverly, MA).
8. 1X T4 DNA polymerase buffer: 50 m*M* NaCl, 10 m*M* Tris-HCl, pH 7.9, 10 m*M* $MgCl_2$, 1 m*M* dithiothreitol (DTT), 50 μg/mL BSA.

9. T4 DNA ligase (New England Biolabs, Beverly, MA).
10. 1X T4 DNA ligase buffer: 50 m*M* Tris-HCl, pH 7.5, 10 m*M* $MgCl_2$, 10 m*M* DTT, 1 m*M* ATP, 25 µg/mL BSA.
11. *Sma*I, *Hind*III, *Pvu*II, *BamH*I restriction enzymes (New England Biolabs, Beverly, MA).
12. Agarose gel equipment.
13. LB (Luria-Bertani) medium: 1% bacto tryptone, 0.5 % yeast extract, 1% NaCl.
14. SOB medium: 2% bacto tryptone, 0.5% yeast extract, 10 m*M* NaCl, 2.5 m*M* KCl, 10 m*M* $MgCl_2$, 10 m*M* $MgSO_4$.
15. SOC medium: 2% bacto tryptone, 0.5% yeast extract, 10 m*M* NaCl, 2.5 m*M* KCl, 10 m*M* $MgCl_2$, 10 m*M* $MgSO_4$, 20 m*M* glucose.
16. FB buffer: 100 m*M* KCl, 50 m*M* $CaCl_2 \cdot 2H_2O$, 10% glycerol, 10 m*M* MOPS, pH 6.5.
17. Kanamycin.
18. Rifampicin.
19. Streptomycin.
20. DMF (dimethylformamide).
21. pCR-Blunt vector (Invitrogen, Carlsbad, CA).
22. pMAK705 (S. R. Kushner, University of Georgia, GA).
23. Spectrophotometer.
24. QIAprep Spin Miniprep kit (QIAGEN, Valencia, CA).

3. Methods

3.1. Mutator Plasmid

pMutD5 plasmid consists of 3 components: 1) temperature-sensitive origin of replication derived from a variant of pSC101 plasmid, pMAK705 *(5)*; 2) *mutD5* gene obtained from genomic DNA of *E. coli* CSH116 *(6)*; 3) Kanamycin resistance marker from pCR-Blunt vector.

1. Amplify *mutD5* gene by PCR using primers mutd1 (5'-CGCCTCCAGCGCGACAATAGCGGCCATC-3') and mutd2 (5'-CCGACTGAACTACCGCTCCGCGTTGTG-3').
2. Clone the PCR product into the pCR-Blunt vector.
3. Check the orientation of *mutD5* insert by digesting the obtained plasmids with *BamH*I. Expect to see 2 bands on an agarose gel stained with ethidium bromide: 0.45 kb fragment carrying the portion of the vector and the beginning of the *mutD5* gene and 4.2 kb fragment carrying the rest of the gene and the vector.
4. Digest plasmid containing the *mutD5* insert in the same orientation as kanamycin resistance marker with *Sma*I and *Hind*III and fill *Hind*III overhangs using T4 polymerase.
5. Clone the DNA fragment carrying *mutD5* gene and kanamycin marker into pMAK705 plasmid digested with *Sma*I and *Pvu*II.
6. Transform the ligation product into *E. coli* competent cells according to manufacturer's instructions. The resulting plasmid will contain the temperature-sensitive origin of replication, *mutD5* gene and kanamycin resistance marker (*see* **Fig. 2**) (*see* **Note 1**).

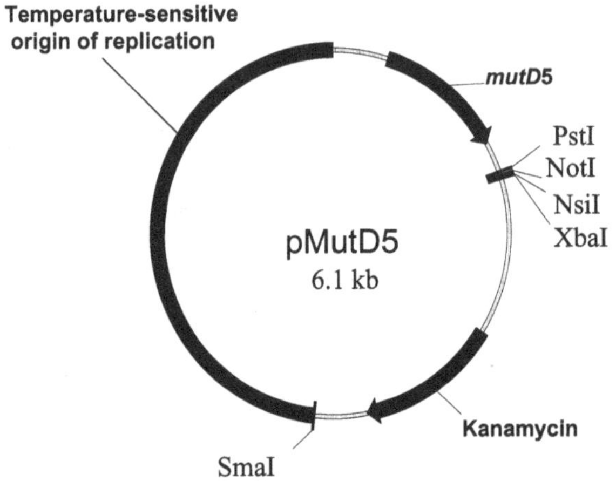

Fig. 2. Schematic drawing of the mutator plasmid pMutD5.

3.2. Strain Mutagenesis and Variant Selection

3.2.1. Preparation of Chemically Competent Cells (7,8)

1. Inoculate 1 fresh colony into 10-mL SOB medium in 125 ml flask and grow at 37°C to $OD_{550}=0.3-0.5$.
2. Transfer cell suspension to a 50-mL Falcon 2070 centrifuge tube and chill on ice for 10–60 min.
3. Centrifuge the cells at 1500*g* for 15 min at 4°C and remove the supernatant. Try to remove the last drops with a paper towel.
4. Gently resuspend cells in 3.3 mL FB buffer and incubate on ice for 10–60 min.
5. Centrifuge the cells at 1500*g* for 15 min at 4°C and remove the supernatant.
6. Gently resuspend cells in 0.8 mL FB buffer.
7. Dispense competent cells in 200-µL aliquots in 2-mL plastic tubes and freeze at –70°C.

3.2.2. Transformation of Competent Cells with pMutD5

1. Add 2–5 µL pMutD5 plasmid DNA to chemically competent cells and incubate on ice for 10–60 min.
2. Heat-shock the cells for 90 s at 42°C.
3. Quickly transfer the tube to ice.
4. Add 800 µL SOC medium and incubate at 37°C for 30–60 min.
5. Spread different volumes of cells on selective plates and incubate overnight.

3.2.3. Testing for Mutation Frequency

1. Grow the cells carrying mutator plasmid in liquid LB medium containing 25 µg/mL kanamycin to an OD_{600} of approx 1.0.

2. Determine the mutator phenotype by plating 10, 50, and 100 μL of cell suspension on LB plates with 100 μg/mL rifampicin (*E. coli*) or 100 μg/mL streptomycin (*E. blattae*) ***(9)***.
3. Dilute the cell suspension 10-, 100-, 1000-, 10,000-fold and plate on LB without antibiotic for calculation of the total cell number.
4. Next day calculate the mutation frequency by dividing the number of resistant colonies by the total number of plated cells (*see* **Notes 2** and **3**).

3.2.4. Evolution Using Mutator Plasmid

1. Grow cells harboring the mutator plasmid under appropriate selection conditions to achieve desired phenotype.

3.2.5. Example of Solvent Tolerance Evolution

Evolution of solvent tolerance was performed on LB plates supplemented with DMF at 50, 60, 70, 80, and 90 g/L and with kanamycin at 25 μg/mL. Since pMutD5 plasmid provides a very high mutation rate, the size of evolving population was limited to 10^6 cells per plate. Raised colonies were counted after 3 d of growth and the 10 largest colonies were selected for the next plating. The cells of the colonies were mixed together, and after the OD measurement, fresh samples containing 10^6 cells were plated on new plates containing the same or higher concentration of DMF. The same cultures were tested on rifampicin plates to monitor the mutator phenotype. Two consecutive rounds of selection were sufficient to obtain the desired improvement in solvent tolerance. Evolved MM294 and *E. blattae* 33429 grew in the presence of 80 g/L and 70 g/L DMF respectively. Wild-type cultures of *E. coli* MM294 and *E. blattae* 33429 grew in the presence of DMF at 60 g/L and 50 g/L respectively (*see* **Notes 4–6**).

3.3. Stabilization of Improved Variants

3.3.1. Plasmid Curing

1. Grow *E. coli* cells carrying pMutD5 plasmid in 5 mL LB medium without antibiotic at 43°C overnight.
2. Transfer 50 μL of cell suspension into 5 mL fresh LB and incubate at 43°C for 8 h.
3. Transfer 50 μL of cell suspension into 5 mL fresh LB and incubate at 43°C overnight.
4. Dilute cell suspension 100, 1000, 10,000-fold and plate small aliquots to obtain single colonies.
5. Transfer cells of individual colonies in parallel fashion on nonselective (LB) and selective (LB supplemented with 25 μg/mL kanamycin) plates.
6. Collect clones that have lost ability to grow on kanamycin and analyze by standard plasmid purification using QIAprep Spin Miniprep kit according to manufacturer's instructions.

Colonies that do not grow on kanamycin and do not show any plasmids on the agarose gel are considered cured (*see* **Note 7**).

3.3.2. Characterization of Winners

1. Grow cured clones under final selection conditions that were used before curing.
2. Make 2–3 transfers at the same conditions to determine stability of evolved features.

3.3.3. Example of Characterization of DMF Tolerant Mutants

Cured clones were tested for newly evolved traits on LB plates or in liquid medium supplemented with varying DMF concentrations. Evolved cultures of *E. coli* MM294 and the control wild-type cultures were tested in liquid LB medium containing 0, 60, 70, or 80 g/L DMF. Initially, growth was observed in all cultures after overnight incubation, however, after the second transfer only cultures evolved with mutator plasmid were able to grow with DMF at 80 g/L. It should be noted that during the curing procedure, the cells were grown in the absence of DMF for more that 30 doublings. These results show that the selected phenotype is stable even in the absence of any selection pressure.

4. Notes

1. pMutD5 plasmid has been shown to be effective for evolution of solvent resistance ***(3)***. The temperature-sensitive origin of replication allows to cure the cells of the plasmid rapidly by growth at elevated temperature (43–44°C) following 1–2 consecutive transfers of the cell culture. Other plasmid systems like pBR322 can also be used for cloning *mutD5* gene and subsequently for evolution of bacterial cultures. However, with pBR322, more time is required for curing the cells of the plasmid.
2. Total number of plated cells was determined by plating various dilutions of the culture on LB plates without antibiotic. The approximate number of cells was calculated using the following assumptions: $OD_{600} = 1.0$ represents approx 2.5×10^8 *E. coli* cells per 1 mL cell culture. Actual OD_{600} number allows to determine the amount of cells per mL. It is important to dilute the cell suspension 10-fold for accurate OD measurement. Turbidity measurement is not adequate in any spectrophotometer after it reaches an OD_{600} value of more than 0.5–0.7.
3. The mutation frequency can be adjusted by altering the expression level of the *mutD5* allele. This can be done by changing the start codon of the *mutD5* gene from ATG to TTG or GTG or by altering the promoter strength.
4. It is important to minimize culturing of any mutator strain to avoid accumulation of undesired mutations. A strain carrying the *mutD5* allele has a rate of spontaneous mutagenesis that is up to 10^3 times higher than a wild type strain. Wild type *E. coli* has a mutation rate of 5×10^{-10} mutations per doubling per base pair ***(10)***. If the genome of *E. coli* is considered to have 4.6×10^6 base pairs, then a *mutD5* strain will acquire on an average more than two mutations

per doubling per genome. Although, selection during growth counteracts the introduction of detrimental mutations, mutator strains will accumulate mutations that are neutral under the current culture conditions but that can be detrimental under other conditions ***(11)***.

5. The mutation rate of mutD5 strains depends on the culture conditions. In rich medium, the mutation rate of mutD5 strains is about 1000 to 10,000-fold elevated compared to wild type *E. coli*. In defined medium, the mutation rate is elevated by only 50 to 100-fold ***(12)***. The mutation rate appears to be dependent on the growth rate, temperature, and nutrient conditions. Thus it may be important to test the mutation rate of a strain under the given selection conditions.
6. The mutator plasmid described here was shown to function in *E. blattae*. Its effectiveness in other organisms requires sufficient expression of the *mutD5* protein and its effective binding to the DNA polymerase. Some organisms like *Bacillus* express the DNA polymerase and its proofreading domain as a single protein. In such cases it should be possible to clone the non-proofreading allele of the entire protein on a plasmid.
7. Curing of different bacterial strains of regular plasmids can be assisted using various agents like mitomycin C, ethidium bromide, sodium dodecyl sulphate, elevated temperature, or high-voltage electroporation ***(13,14)***. After subculturing the cells in the absence of antibiotics for several generations, 100 random colonies are plated in parallel on nonselective and selective media. The efficiency of curing is lower for regular plasmids compared to plasmids with temperature-sensitive origin of replication. There are rare cases when even mutator plasmids carrying temperature-sensitive origin of replication cannot be cured. In this study, *E. coli* W1485 strain could not be cured after it was evolved to grow with 70 g/L DMF. The mutator plasmid was curable from W1485 strain prior to evolution of DMF tolerance. High-molecular-weight multimers of pMutD5 were detected in evolved W1485 strain that were not lost even after prolonged culturing at elevated temperature. The lack of curing could be associated with specific traits of a strain. Thus, it is imperative to test several related bacterial strains for such evolution experiments.
8. If directed evolution is performed with very large populations, then the spontaneous mutation rate of microorganisms may be sufficient to provide variants with novel phenotypes. However, it has been observed on many occasions that mutators strains tend to arise spontaneously when large populations of bacteria are placed under prolonged selection pressure ***(15,16)***. It is difficult to stabilize such spontaneous mutator strains as one first needs to determine the defect that caused the mutator phenotype. By introducing the mutator plasmid one eliminates the selection pressure that normally leads to the accumulation of spontaneous mutator strains.

Acknowledgments

We thank Dr. Fernando Valle for construction of the pMutD5 plasmid and Roopa Ghirnikar for help with the preparation of manuscript. We would like to thank Dr. Sidney Kushner and Valerie Maples, University of Georgia, Athens, for the sequence of plasmid pMAK705.

References

1. Roa, A., Garcia, J. L., Salto, F., and Cortes, E. (1994) Changing the substrate specificity of penicillin G acylase from *Kluyvera citrophila* through selective pressure. *Biochem. J.* **303,** 869–876.
2. Long-McGie, J., Liu, A. D., and Schellenberger, V. (2000) Rapid in vivo evolution of a β-lactamase using phagemids. *Biotechnol. Bioeng.* **68,** 121–125.
3. Selifonova, O., Valle, F., and Schellenberger, V. (2001) Rapid evolution of novel traits in microorganisms. *Appl Environ. Microbiol.* **67,** 3645–3649.
4. Cox, E. C. and Horner, D. L. (1983) Structure and coding properties of a dominant *Escherichia coli* mutator gene, mutD. *Proc. Natl. Acad. Sci. USA* **80,** 2295–2299.
5. Hamilton, C. M., Aldea, M., Washburn, B. K., Babitzke, P., and Kushner, S. R. (1989) New method for generating deletions and gene replacements in *Escherichia coli. J. Bacteriol.* **171,** 4617–4622.
6. Miller, J. H. (1992) *A Short Course in Bacterial Genetics*, Cold Spring Harbor Press, Cold Spring Harbor, NY.
7. Hanahan, D. (1985) Techniques for transformation of *E. coli*, in *DNA Cloning: A Practical Approach, vol. 1*, (Glover, D. M., ed.), IRL Press, Oxford, Washington DC, pp. 109–136.
8. Sambrook, J., Fritsch, E. F., and Maniatis, T. (1989) *Molecular Cloning: A Laboratory Manual*, Cold Spring Harbor Press, Cold Spring Harbor, NY.
9. Horiuchi, T., Maki, H., and Sekiguchi, M. (1978) A new conditional lethal mutator (dnaQ49) in *Escherichia coli* K12. *Mol. Gen. Genet.* **163,** 227–283.
10. Drake, J. W. (1991) A constant rate of spontaneous mutation in DNA-based microbes. *Proc. Natl. Acad. Sci. USA* **88,** 7160–7164.
11. Giraud, A., Matic, I., Tenaillon, O., et al. (2001) Costs and benefits of high mutation rates: adaptive evolution of bacteria in the mouse gut. *Science* **291,** 2606–2608.
12. Degnen, G. E. and Cox, E. C. (1974) Conditional mutator gene in *Escherichia coli*: Isolation, mapping, and effector studies. *J. Bacteriol.* **117,** 477–487.
13. El-Mansi, M., Anderson, K. J., Inche, C. A., Knowles, L. K., and Platt, D. J. (2000) Isolation and curing of the *Klebsiella pneumoniae* large indigenous plasmid using sodium dodecyl sulphate. *Res. Microbiol.* **151,** 201–208.
14. Heery, D. M., Powell, R., Gannon, F., and Dunican, L. K. (1989) Curing of a plasmid from *E. coli* using high-voltage electroporation. *Nucleic Acids Res.* **17,** 10,131.
15. Mao, E. F., Lane, L., Lee, J., and Miller, J. H. (1997) Proliferation of mutators in a cell population. *J. Bacteriol.* **179,** 417–422.
16. Sniegowski, P. D., Gerrish, P. J., and Lenski, R. E. (1997) Evolution of high mutation rates in experimental populations of *E. coli. Nature* **387,** 703–705.

8

Random Insertion and Deletion Mutagenesis

Hiroshi Murakami, Takahiro Hohsaka, and Masahiko Sisido

1. Introduction

Random mutagenesis combined with high-throughput screening is a versatile strategy for improving protein functions or creating artificial enzymes *(1,2)*. Several methods for introducing random mutations in vitro have been reported *(3)*. Among these, error-prone PCR mutagenesis, based on inaccurate copying by DNA polymerase, is the most commonly used technique to introduce random point mutations *(4)*. However, the error-prone PCR method has an inherent drawback of biased occurrence of amino acids as the result of single base replacements in the triplet codons. For example, mutation from AUG (Met) to UGG (Trp) is unlikely to take place. To achieve a non-biased random replacement on the amino acid level, oligonucleotide-directed mutagenesis *(5)* and cassette mutagenesis *(6,7)* have been carried out. These methods are limited to a defined region of the gene and can not introduce mutations at random positions. To introduce codon-based mutations at various positions, the split-and-mix method *(8)* or other synthetic methods of constructing DNAs using dinucleotide or trinucleotide units have been attempted *(9–11)*. These synthetic methods may be applicable only to introduce mutations within a narrow range of the target gene.

We have developed a novel method to introduce codon based mutations at random positions along the entire range of the target gene (**Fig. 1**; *12*). The method, random insertion/deletion (RID) mutagenesis, enables deletion of an arbitrary number of consecutive bases at random positions and, at the same time, insertion of a specific sequence or random sequence of an arbitrary number of bases into the same position. Using this method, we can replace three randomly selected consecutive bases by a specific codon, by a mixture of codons for different amino acids, or even by four-base codons for nonnatural amino acids *(13,14)*.

From: *Methods in Molecular Biology, vol. 231: Directed Evolution Library Creation: Methods and Protocols*
Edited by: F. H. Arnold and G. Georgiou © Humana Press Inc., Totowa, NJ

wild-type (GFP$_{UV}$)
TTC TCT TAT
64F 65S 66Y
⇩
TTA ACT TAT
64L 65T 66Y
mutant (F64L, S65T)

wild-type (GFP$_{UV}$)
TCG ACA CAA
202S 203T 204Q
⇩
TCG TAT CAA
202S 203Y 204Q
mutant (T203Y)

Fig. 1. A general scheme of the random insertion/deletion (RID) mutagenesis for the construction of a library of mutant genes. The procedure consists of eight major steps that are described in **Subheading 3.4.** (1) The fragment obtained by digesting the original gene with *Eco*RI and *Hin*dIII is ligated with a linker. Then the product is digested with *Hin*dIII to make a linear dsDNA with a nick in the antisense chain. (2) The gene fragment is cyclized with T4 DNA ligase to make a circular dsDNA with a nick in the antisense chain. (3) The circular dsDNA is treated with T4 DNA polymerase to produce a circular ssDNA. (4) The circular ssDNA is randomly cleaved at single positions by treating with Ce(IV)-EDTA complex. (5) The linear ssDNAs with unknown sequences at both ends are ligated to the 5'-anchor and the 3'-anchor, respectively. (6) The DNAs linked with the two anchors at both ends are amplified by PCR. (7) The PCR products are treated with *Bci*VI, leaving several bases that have been included in the 5'-anchor to the one end. The *Bci*VI treatment also deletes a specific number of bases from the other end. (8) The digested products are treated with Klenow fragment to make blunt ends and cyclized again with T4 DNA ligase. The products are treated with *Eco*RI and *Hin*dIII, and the fragments obtained are cloned into an *Eco*RI-*Hin*dIII site of modified pUC18 (pUM).

The applicability of the RID mutagenesis was demonstrated by random replacement of three consecutive bases by a recognition sequence of *Bgl*II (AGATCT) in the GFP$_{UV}$ gene. Alternatively, three randomly selected bases were replaced by a mixture of 20 codons of 20 naturally-occurring amino acids in the GFP$_{UV}$ gene. The mutants were expressed in *E. coli*, and the yellow fluorescent protein and the enhanced green fluorescent protein that could not be obtained by the conventional error-prone PCR mutagenesis were found.

2. Materials

1. pUC18 (Takara, Japan).
2. pGFP$_{UV}$ (Clontech, Palo Alto, CA).
3. Oligonucleotides (Life Technologies, Rockville, MD).
4. Fluorescence labeled oligonucleotide (Amersham Pharmacia Biotech, Piscataway, NJ).
5. *Eco*RI, *Hin*dIII, *Bgl*II, *Bci*VI (New England BioLabs, Beverly, MA).
6. T4 DNA ligase, T4 polynucleotide kinase, T4 DNA polymerase, Klenow fragment (New England BioLabs).
7. Pfx DNA polymerase (Life Technologies).

8. Glycogen (20 mg/mL).
9. QIA quick PCR purification kit (QIAGEN, Valencia, CA).
10. DE-52 cellulose (Whatman, England).
11. Ceric ammonium nitrate.
12. Streptavidin-agarose (PIERCE).
13. Fluorescence image analyzer (Hitachi FM-BIO II instrument).
14. Sharp cut filters (Toshiba, L44 and Y52).
15. *E. coli* strain DH10B (for electroporation) and DH5α.
16. 0.1 *M* IPTG (isopropyl-β-D-thio-galactopyranoside).
17. 100 mg/mL Amp (ampicillin).
18. T4 DNA polymerase buffer: 20 m*M* Tris-acetate, pH 7.9, 50 m*M* potassium acetate, 10 m*M* magnesium acetate, 1 m*M* dithiothreitol (DTT), 100 μg/mL bovine serum ablumin (BSA).
19. T4 Polynucleotide kinase buffer: 70 m*M* Tris-HCl, pH 7.6, 10 m*M* $MgCl_2$, 5 m*M* dithiothreitol, 1 m*M* ATP.
20. *Bci*VI buffer: 20 m*M* Tris-acetate, pH 7.9, 50 m*M* potassium acetate, 10 m*M* magnesium acetate, 1 m*M* dithiothreitol.
21. Klenow fragment buffer: 10 m*M* Tris-HCl, pH 7.5, 5 m*M* $MgCl_2$, 7.5 m*M* dithiothreitol, 33 μ*M* dNTPs.
22. Ligation buffer A: 25 m*M* Tris-HCl, pH 7.5, 8% PEG8000, 0.5 m*M* ATP, 5 m*M* dithiothreitol, 5 m*M* $MgCl_2$, and 12 μg/mL BSA.
23. Ligation buffer B: 50 m*M* Tris-HCl, pH 7.5, 8% PEG8000, 10 m*M* $MgCl_2$, 10 m*M* dithiothreitol, 1 m*M* ATP, 25 μg/mL BSA, and 150 m*M* NaCl.
24. Wash buffer A: 10 m*M* Tris-HCl, pH 7.0, 50 m*M* phosphate, 1 m*M* ethylenediamine tetraacetic acid (EDTA), 100 m*M* NaCl.
25. Wash buffer B: 10 m*M* Tris-HCl, pH 7.0, 1 m*M* EDTA, 100 m*M* NaCl.
26. Elution buffer C: 4 parts of 10 m*M* Tris-HCl, pH 7.0, 1 m*M* EDTA, 1.5 *M* NaCl, and 6 parts of formamide.
27. 10 m*M* Tris-HCl, pH 7.0, 1 m*M* EDTA, 1 *M* NaCl.
28. 50 m*M* HEPES-NaOH, pH 7.5.
29. 1 *M* Tris-HCl, pH 7.0.
30. 0.11 *M* HEPES-NaOH, pH 7.5, and 11 m*M* EDTA.
31. 2 *M* NaOH.
32. 0.25 *M* phosphate, pH 7.0, 0.25 *M* NaCl.

3. Methods

3.1. Plasmid Construction (see Note 1)

Plasmid (pUM-GFP_{UV}) was obtained by the following procedure. The pUM fragment was amplified from pUC18 by PCR using 5'-AAATAAGCT TGGCACTGGCCGTC-3' and 5'-CATGAATTCCGTAATCATGGTCAT AGCTG-3' as primers with Pfx DNA polymerase. The DNA fragment of GFP_{UV}-coding region was amplified from pGFP_{UV} *(15)* by PCR using 5'-GGAATTCATGAGTAAAGGAGAAGAACT-3' and 5'-GCCCAAGCTTAT TTGTAGAGCTCATCCA-3' as primers to introduce restriction sites. These

A

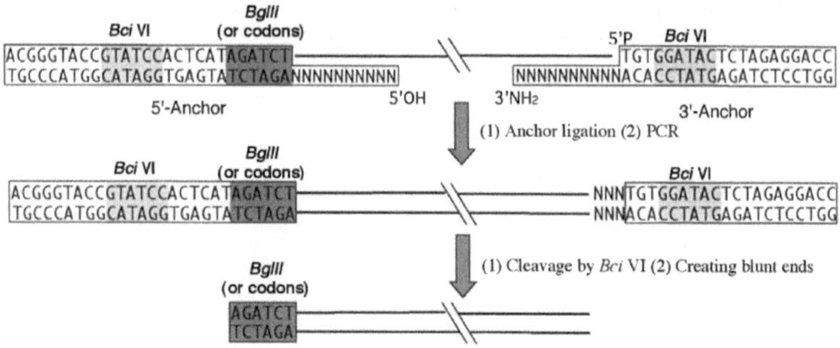

B

Gly	GGC	Ser	TCT	Arg	CGC	Phe	TTT
Ala	GCC	Thr	ACC	Asn	AAC	Tyr	TAT
Val	GTG	Cys	TGC	Gln	CAG	Trp	TGG
Leu	CTG	Lys	AAA	Asp	GAT	Pro	CCG
Ile	ATT	His	CAT	Glu	GAA	Met	ATG

Fig. 2. Sequences of the 5'- and 3'-anchors. **(A)** The 5'-anchor contains *Bci*VI (GTATCC) and *Bgl*II (AGATCT) recognition sequences and a 10-base tail of random sequence for hybridization to the 5'-end of linear ssDNA with unknown sequence. The 3'-anchor contains a *Bci*VI recognition sequence and a 10-base random tail. When the anchor-linked DNA is amplified by PCR, treated with *Bc*NI, and then treated with Klenow fragment, the AGATCT sequence will be left on one end and the terminal three bases will be deleted from the other end. **(B)** Twenty codons that direct 20 naturally-occurring amino acids were introduced into the GFP_{UV} gene by RID mutagenesis. The codons were put into the 5'-anchor in place of the *Bgl*II recognition sequence.

fragments were treated with *Eco*RI and *Hin*dIII. The fragment of GFP_{UV}-coding region was cloned into the *Eco*RI and *Hin*dIII site of pUM. DH5α *E. coli* cells were transformed with pUM-GFP_{UV} using standard molecular biology methods.

3.2. Preparation of Linker and Anchors

The linker used in **Subheading 3.4.1.** was prepared by annealing the oligonucleotides, 5'-CAGTCGCAAGCTTGGCATGGTGGG-3' and 5'-AATTCC CACCATGCCAAGCTTGCGACTG-3'. The fluorescence labeled linker used for product analysis was also prepared by annealing the oligonucleotides, 5'-FAM-AGCTATGACCATGATTACGG-3' and 5'-AATTCCGTAATCATG GTCATAGCT-3'. The 5'-anchors and 3'-anchor used in **Subheading 3.4.5.** were prepared in a similar manner. The nucleotide sequences of the anchors are shown in **Fig. 2**. For random mutation experiment, all 20 anchors were mixed after annealing individual sense and antisense pairs.

1. Mix 25 μL of the sense chain and 25 μL of the antisense chain in 0.5 μL of 1 *M* Tris-HCl, pH 7.0 (final concentration: 100 μ*M* each).
2. Incubate the mixture at 90°C for 30 sec, 50°C for 2 min, 37°C for 2 min. Store at –20°C.

3.3. Preparation of Ce-EDTA Complex (see Note 2)

1. Prepare 0.1 *M* ceric ammonium nitrate in water.
2. Dilute 10 fold with 0.11 *M* HEPES-NaOH, pH 7.5, that contained 11 m*M* EDTA.
3. Adjust the pH to 7.5 by adding 2 *M* NaOH.
4. Dilute the Ce-EDTA solution to 0.6 m*M* with 50 m*M* HEPES-NaOH, pH 7.5.

3.4. RID Mutagenesis

3.4.1. Preparation of DNA Fragment of Target Gene (see **Note 3**)

1. Grow DH5α *E. coli* cells transformed with pUM-GFP_{UV} in 500 mL of 2X TY media containing ampicillin, overnight at 37°C.
2. Purify the plasmid DNA by standard methods.
3. Cleave 1 mg of plasmid with *Eco*RI and *Hin*dIII, purify the DNA fragment by a standard method using polyacrylamide gel electrophoresis (PAGE). Dissolve the DNA in water.
4. Prepare ligation solution from 200 μL of the ligation buffer A, 1.2 nmol of linker and 400 U of T4 DNA ligase.
5. Add 6 μg (12 pmol) of DNA into 200 μL of the ligation solution, incubate 10 min at 37°C, and repeat the addition-incubation procedure for further 4 × until a total of 30 μg (60 pmol) has been added.
6. Extract the reaction mixture with phenol/chloroform (50:50) and with chloroform. Precipitate the product with glycogen (10 μg) and 0.6 vol 2-propanol. Spin the mixture for 30 min and wash the precipitate with 70% ethanol. Dry the precipitate and dissolve in water.
7. Cleave the DNA with *Hin*dIII by standard methods, extract the reaction mixture with phenol/chloroform (50:50) and with chloroform. Precipitate the product with 0.1 vol of 3 *M* AcONa and 0.6 vol of 2-propanol. Spin the mixture for 30 min and wash the precipitate with 70% ethanol. Dry the precipitate and dissolve in water.

3.4.2. Cyclization of DNA Fragment (see **Notes 4** and **5**)

1. Prepare a ligation solution from 6 mL of the ligation buffer A and 500 U of T4 DNA ligase.
2. Add 6 μg (12 pmol) of the DNA into 6 mL of ligation solution, incubate 5 min at 16°C, and repeat the addition-incubation procedure for further 4 × until a total of 30 μg (60 pmol) has been added.
3. Extract the reaction mixture with phenol/chloroform (50:50) and with chloroform. Precipitate the product with 0.1 vol 3 *M* AcONa and 2.5 vol ethanol. Spin the mixture for 30 min and wash the precipitate with 70% ethanol. Dry the precipitate and dissolve in water.

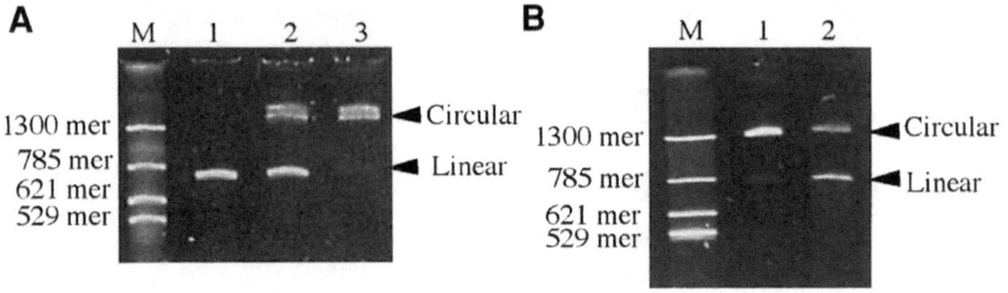

Fig. 3. Analysis of the products in **Subheading 3.4.1. to 3.4.4.** on 8 *M* Urea/4% PAGE. Lane *M* shows DNA size marker. **(A)** Analysis of the products in **Subheading 3.4.1. to 3.4.4.**; Lane 1, dsDNA fragment obtained in **Subheading 3.4.1.**; Lane 2, products of the cyclization in **Subheading 3.4.2.**; Lane 3, products after treatment with T4 DNA polymerase in **Subheading 3.4.3. (B)** Analysis of the products of **Subheading 3.4.4.** Lane 1, before the Ce(IV)-EDTA treatment (same as Lane 3 in **A**); Lane 2, After the Ce(IV)-EDTA treatment in **Subheading 3.4.4.**

*3.4.3. Remove Antisense Chain (see **Note 6**)*

1. Add the DNA (50 pmol, 25 μg) to 100 μL of T4 DNA polymerase buffer, which contains 15 units of T4 DNA polymerase and incubate at 37°C for 2 h.
2. Purify the product using a QIA quick PCR purification kit.

*3.4.4. Random Cleavage of the Circular ssDNA (see **Notes 7–9**)*

1. Mix 20 μL of DNA fragment solution (4 pmol, 1 μg), 70 μL of water and 10 μL of 500 m*M* HEPES-NaOH, pH 7.5.
2. After heating the DNA solution at 94°C for 30 sec, add 100 μL of 0.6 m*M* Ce-EDTA solution (*see* **Subheading 3.3.**), and heat the reaction mixture at 94°C for 30 sec. Quench the reaction by adding 20 μL of 0.25 *M* phosphate, pH 7.0, 0.25 *M* NaCl.
3. Add the reaction mixture to DE-52 column (10 μL), wash with 400 μL of wash buffer A and with 50 μL of wash buffer B.
4. Elute DNA with 50 μL of elution buffer C by heating to 80°C for 1 min.
5. Add 130 μL of water and precipitate the product with 0.1 vol of 3 *M* AcONa and 2.5 vol of ethanol. Spin the mixture for 30 min and wash the precipitate with 70% ethanol. Dry the precipitate and dissolve in water (*see* **Fig. 3**).

*3.4.5. Anchor Ligation (see **Notes 10–12**)*

1. Add the DNA (4 pmol, 1 μg) to 20 μL of T4 polynucleotide kinase buffer.
2. Heat the solution at 90°C for 20 sec and then place on ice.
3. Add 4 U of T4 polynucleotide kinase and incubate the mixture at 37°C for 10 min.
4. Extract the reaction mixture with phenol/chloroform (50:50) and with chloroform. Precipitate the product with 0.1 vol of 3 *M* AcONa and 2.5 vol of ethanol.

Spin the mixture for 30 min and wash the precipitate with 70% ethanol. Dry the precipitate and dissolve in water.

5. Add the DNA (0.8 pmol, 200 ng) to 10 μL of ligation solution B, which contains 50 pmol of each 5'-anchor and 3'-anchor and 400 units of T4 DNA ligase.
6. Incubate for 12 h by cycling temperatures between 10°C and 30°C at 12 cycles/h ***(16)***.
7. Extract the reaction mixture with phenol/chloroform (50:50) and with chloroform. Precipitate the product with 0.1 vol of 3 *M* AcONa and 2.5 vol of ethanol. Spin the mixture for 30 min and wash the precipitate with 70% ethanol. Dry the precipitate and dissolve in water.
8. Purify the DNA by 4% polyacrylamide/8 *M* urea gel.

3.4.6. PCR Amplification of DNA Fragment Ligated with the Two Anchors

1. Add the purified DNA to 200 μL of PCR solution, which contains biotin-5'-ACGGGTACCGTATCCACTCA-3' and biotin-5'-GGTCCTCTAGAGTATCCACA-3' as primers and 1.5 U of Pfx DNA polymerase.
2. Amplify DNA by temperature cycling (94°C 20 sec, 50°C 30 sec, 68°C 50 sec). Stop the reaction while the amplification is still in the log-phase (12–18 cycles).
3. Purify the PCR product using a QIA quick PCR purification kit.

*3.4.7. Cleavage of Both Ends by BciVI and Removal of 3'-Overhang (see **Note 13**)*

1. Add the PCR product (0.4 pmol, 200 ng) to 5 μL of *Bci*VI buffer which contains 5 units of *Bci*VI and incubate at 37°C for 90 min.
2. Mix the reaction mixture with 20 μL of 10 m*M* Tris-HCl, pH 7.0, 1 m*M* EDTA, 1 *M* NaCl, and 20 μL of streptavidin-agarose.
3. Incubate the solution at room temperature for 30 min with moderate agitation.
4. Add the solution to an empty column and collect the solution phase.
5. Extract the reaction mixture with phenol/chloroform (50:50) and with chloroform. Precipitate the product with 0.1 vol of 3 *M* AcONa and 2.5 vol of ethanol. Spin the mixture for 30 min and wash the precipitate with 70% ethanol. Dry the precipitate and dissolve in water.
6. Add the DNA to 10 μL of Klenow fragment buffer which contains 0.5 units of Klenow fragment, and incubate the mixture at 25°C for 15 min.
7. Extract the reaction mixture with phenol/chloroform (50:50) and with chloroform. Precipitate the product with 0.1 vol of 3 *M* AcONa and 2.5 vol of ethanol. Spin the mixture for 30 min and wash the precipitate with 70% ethanol. Dry the precipitate and dissolve in water.

3.4.8. Reconstitution of Target Gene and Cloning

1. Add the DNA to 150 μL of ligation solution A, which contains 600 U of T4 DNA ligase. Incubate the mixture at 16°C for 2 h.

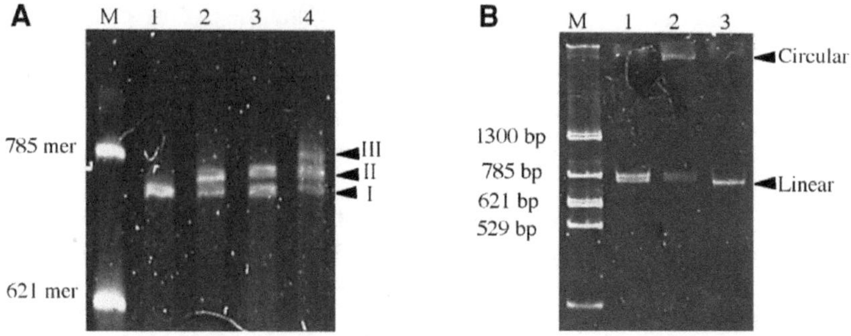

Fig. 4. Analysis of the products in **Subheading 3.4.5.** on 8 *M* Urea/4% PAGE and **Subheadings 3.4.6. to 3.4.8.** on 4% PAGE. (**A**) Analysis of the ligation products of **Subheading 3.4.5.** Lane 1, ligation products without anchors; Lane 2, ligation products in the presence of 3'-anchor; Lane 3, ligation products in the presence of 5'-anchor that contains the *Bgl*II site. Lane 4, ligation products in the presence of both 3'-anchor and 5'-anchor (*Bgl*II). I; ssDNA without the anchor ligation, II; ssDNA linked with the anchor at one of the two ends, III ssDNA linked with the anchors at both ends. (**B**) Analysis of the products in **Subheadings 3.4.6. to 3.4.8.** Lane 1, before the cyclization; Lane 2, after the cyclization; Lane 3, after the *Eco*RI and *Hin*dIII cleavage of the cyclization product.

2. Extract the reaction mixture with phenol/chloroform (50:50) and with chloroform. Precipitate the product with 0.1 vol of 3 *M* AcONa and 2.5 vol ethanol. Spin the mixture for 30 min and wash the precipitate with 70% ethanol. Dry the precipitate and dissolve in water.
3. Cleave the DNA with *Hin*dIII and *Eco*RI by standard methods. Extract the reaction mixture with phenol/chloroform (50:50) and with chloroform. Precipitate the product with 0.1 vol of 3 *M* AcONa and 0.6 vol of 2-propanol. Spin the mixture for 30 min and wash the precipitate with 70% ethanol. Dry the precipitate and dissolve in water.
4. Insert the DNA fragment into pUM *Eco*RI-*Hin*dIII site and transform DH10B *E. coli* cells by electroporation.
5. Cultivate the cell on LB plates with ampicillin and IPTG.
6. Collect all of cells and purify the plasmid DNA by standard methods (*see* **Fig. 4**).

3.5. Analysis of Mutation Positions

1. Grow DH5α *E. coli* cells transformed with the plasmid library, which was prepared using *Bgl*II anchor, in 12 mL of 2XTY media containing ampicillin overnight at 37°C.
2. Purify the plasmid DNA by standard methods.
3. Cleave 50 μg of the plasmid with *Eco*RI and *Hin*dIII, purify the DNA fragment by PAGE by standard methods, and dissolve in water.

A

Clone	Position	Deletion	Insertion	Point mutation
1	11-13	GAG	AGATCT	
2	31-33	GTT	AGATCT	
3	66-68	TAA	AGATCT	
4	113-115	CAT	AGATCT	
5	123-125	ACT	AGATCT	G100T
6	222-224	TCC	AGATCT	G459T
7	272-274	GTT	AGATCT	
8	317-319	ACA	AGATCT	
9	326-328	GTG	AGATCT	
10	401-403	GAA	AGATCT	C144T
11	405-408	CATT	AGATCT	
12	418-420	AAA	AGATCT	
13	423-425	CGA	AGATCT	G244C
14	512-514	TTG	AGATCT	A539G
15	534-536	AGC	AGATCT	
16	553-555	AAT	AGATCT	G72T
17	588-591	AGAC	AGATCT	
18	693-695	TGG	AGATCT	
19	704-706	AGC	AGATCT	

B

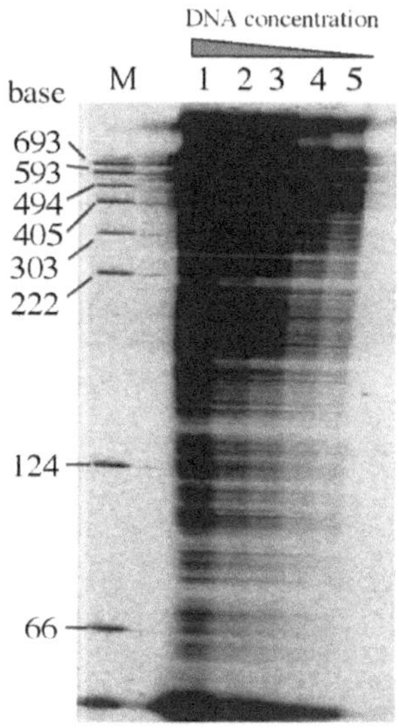

Fig. 5. Sequences of mutant genes and distribution of mutation positions. (**A**) Sequences of mutant genes. The gene library, in which a *Bgl*II recognition sequence (AGATCT) was introduced at random positions, was generated. From the library, 19 randomly selected clones were sequenced and the results are listed. (**B**) The fluorescence labeled linker was attached to the 5'-ends of the sense chain of the mutant genes that contain a *Bgl*II recognition sequence at random positions. The labeled DNAs were digested with *Bgl*II to produce fluorescence labeled fragments of different chain lengths. The products were analyzed on 8 *M* Urea/6% PAGE.

4. Prepare ligation solution from 90 µL of the ligation buffer A, 0.5 nmol of fluorescence labeled linker, and 200 units of T4 DNA ligase.
5. Add 1.2 µg (2.4 pmol) of DNA into 90 µL of the ligation solution, incubate 5 min at 37°C, and repeat the addition-incubation procedure for further 4 × until a total of 6 µg (12 pmol) has been added.
6. Extract the reaction mixture with phenol/chloroform (50:50) and with chloroform. Precipitate the product with glycogen (10 µg) and 0.6 vol 2-propanol. Spin the mixture for 30 min and wash the precipitate with 70% ethanol. Dry the precipitate and dissolve in water
7. Cleave the product with *Bgl*II by standard methods. Extract the reaction mixture with phenol/chloroform (50:50) and with chloroform. Precipitate the product with 0.1 vol of 3 *M* AcONa and 0.6 vol of 2-propanol. Spin the mixture for 30 min and wash the precipitate with 70% ethanol. Dry the precipitate and dissolve in water.
8. Analyze the product on 6% polyacrylamide/8 *M* urea gel electrophoresis recorded on a Hitachi FM-BIO II instrument (*see* **Fig. 5**).

wild-type (GFP_{UV})	wild-type (GFP_{UV})
TTA ACT TAT	TCG ACA CAA
64F 65S 66Y	202S 203T 204Q
⇩	⇩
TTC TCT TAT	TCG TAT CAA
64L 65T 66Y	202S 203Y 204Q
mutant (F64L, S65T)	mutant (T203Y)

Fig. 6. GFP_{UV} mutants obtained by the random substitution with 20 amino acids codons *(20,21)*.

3.6 Colony Selection and Sequence Analysis

The *E. coli* transformed by the plasmid library, which was prepared using random codon anchors, was cultivated on LB plates. The LB plates were irradiated at 400 nm or 490 nm by using a 150 W Xenon lamp equipped with a grating monochrometer. Fluorescence emitted from the colonies on a plate was passed through different sharp cut filters (Toshiba, L44 or Y52), and the colonies that showed intense emission were selected visually (*see* **Fig. 6**).

4. Notes

1. pUM vector was made from pUC18 to adjust reading frame, but it will be possible to use other vectors. If a target gene has a *Bci*VI recognition sequence, it is necessary to introduce a silent mutation to remove the recognition sequence, or use other restriction enzymes such as *Bsg*I or *Bse*RI.
2. The Ce(IV)-EDTA complex tends to precipitate after standing, and resulting in loss of DNA cleavage activity. Use the solution within 2 h after preparation.
3. It is possible to amplify DNA fragments of a target gene by PCR, but it will cause unexpected point mutations, even with high fidelity DNA polymerases such as Pfx polymerase. Preparation of DNA fragment by plasmid digestion is recommended.
4. Use a low concentration of DNA fragment to avoid intermolecular ligation.
5. The high concentration of salt enhances intermolecular ligation rather than intramolecular ligation (cyclization).
6. The nick was introduced in **Subheading 3.4.1.** by using the linker, which has 5'-hydroxyl group at the both ends.
7. Optimize the Ce(IV)-EDTA concentration, if you do not get 50% cleavage efficiency.
8. It is not suitable to use DNase I instead of Ce(IV)-EDTA for cleaving circular DNA. DNase I has some preferred cleavage sites, resulting in a biased library. We strongly recommend use of Ce(IV)-EDTA cleavage. (*see* **ref.** ***17***)
9. Ce(IV)-EDTA cleavage produces almost equal amount of 5'-monophosphate and 3'-monophosphate at both ends (*see* **ref.** ***18***). It is recommended to treat ssDNA products by polynucleotide kinase to increase anchor ligation efficiency in the next step.

10. We have tested 6-, 9-, 10-, 15-mer random tails. The 9- or 10-mer tails gave the best ligation efficiency.
11. The 3'-protection of oligonucleotide was needed to avoid self-ligation of 3'-anchor.
12. This method can insert any short sequence or delete up to a 32-base sequence by using various anchors. For example, alanine scanning mutagenesis ***(19)*** can be achieved by using an insertion anchor, which has an alanine codon sequence. It will be possible to delete 32 bases by removing 16 bases from each end by using *Bsg*I instead of *Bci*VI,
13. T4 DNA polymerase, a high concentration of Klenow fragment, a low concentration of DNA fragments, or long incubation time will cause unexpected deletions. It is very important to use proper conditions to make correct blunt ends.

Acknowledgments

This work was supported by a Grant-in-Aid for Specially Promoted Research from the Ministry of Education, Culture, Sports, Science and Technology, Japan (No.11102003). H. Murakami appreciates the Research Fellowship from the Japan Society for Promotion of Science for Young Scientists.

References

1. Arnold, F. H., Wintrode, P. L., Miyazaki, K., and Gershenson, A. (2001) How enzymes adapt: lessons from directed evolution. *Trends Biochem. Sci.* **26**, 100–106.
2. Brakmann, S. (2001) Discovery of superior enzymes by directed molecular evolution. *Chembiochem.* **2**, 865–871.
3. Braman, J. (ed.) (2001) *In Vitro Mutagenesis Protocols, Second Edition*, Humana, Totowa, NJ
4. Cadwell, R. C. and Joyce, G. F. (1992) Randomization of genes by PCR mutagenesis. *PCR Methods Appl.* **2**, 28–33.
5. Dalbadie-Mcfarland, G., Cohen, L. W., Riggs, A. D., Morin, C., Itakura, K., and Richards, J. H. (1982) Oligonucleotide-directed mutagenesis as a general and powerful method for studies of protein function. *Proc. Natl. Acad. Sci. USA* **79**, 6409–6413.
6. Wells, J. A., Vasser, M., and Powers, D. B. (1985) Cassette mutagenesis: an efficient method for generation of multiple mutations at defined sites. *Gene* **34**, 315–323.
7. Kegler-Ebo, D. M., Docktor, C. M., and Dimaio, D. (1994) Codon cassette mutagenesis: a general method to insert or replace individual codons by using universal mutagenic cassettes. *Nucl. Acids Res.* **22**, 1593–1599.
8. Glaser, S. M., Yelton, D. E., and Huse, W. D. (1992) Antibody engineering by codon-based mutagenesis in a filamentous phage vector system. *J. Immunol.* **149**, 3903–3913.
9. Sondek, J. and Shortle, D. (1992) A general strategy for random insertion and substitution mutagenesis: substoichiometric coupling of trinucleotide phosphoramidites. *Proc. Natl. Acad. Sci. USA* **89**, 3581–3585.

10. Gaytan, P., Yanez, J., Sanchez, F., Mackie, H., and Soberon, X. (1998) Combination of DMT-mononucleotide and Fmoc-trinucleotide phosphoramidites in oligonucleotide synthesis affords an automatable codon-level mutagenesis method. *Chem. Biol.* **5**, 519–527.
11. Neuner, P., Cortese, R., and Monaci, P. (1998) Codon-based mutagenesis using dimer-phosphoramidites. *Nucl. Acids Res.* **26**, 1223–1227.
12. Murakami, H., Hohsaka, T., and Sisido, M. (2002) Random insertion and deletion of arbitrary number of bases for codon-based random mutation of DNAs. *Nat. Biotechnol.* **20**, 76–81.
13. Hohsaka, T., Ashizuka, Y., Taira, H., Murakami, H., and Sisido, M. (2001) Incorporation of nonnatural amino acids into proteins by using various four-base codons in an Escherichia coli in vitro translation system. *Biochemistry* **40**, 11,060–11,064.
14. Hohsaka, T., Ashizuka, Y., Murakami, H., and Sisido, M. (1996) Incorporation of Nonnatural Amino Acids into Streptavidin through In Vitro Frame-Shift Suppression. *J. Am. Chem. Soc.* **118**, 9778–9779.
15. Crameri, A., Whitehorn, E. A., Tate, E., and Stemmer, W. P. (1996) Improved green fluorescent protein by molecular evolution using DNA shuffling. *Nat. Biotechnol.* **14**, 315–319.
16. Lund, A. H., Duch, M., and Pedersen, F. S. (1996) Increased cloning efficiency by temperature-cycle ligation. *Nucl. Acids Res.* **24**, 800-801.
17. Igawa, T. S., Sumaoka, J., and Komiyama, M. (2000) Hydrolysis of oligonucleotides by homogeneous Ce(IV)-EDTA complex. *Chem. Lett.* 56–57.
18. Sumaoka, J. A., Igawa, Y., and Komiyama, M. (1998) Enzymatic manipulation of the fragments obtained by cerium (IV)-induced DNA scission: Characterization of hydrolytic termini. *Chem. Eur. J.* **4**, 205–209.
19. Morrison, K. L. and Weiss, G. A. (2001) Combinatorial alanine-scanning. *Curr. Opin. Chem. Biol.* **5**, 302–307.
20. Cormack, B. P., Valdivia, R. H., and Falkow, S. (1996) FACS-optimized mutants of the green fluorescent protein (GFP). *Gene* **173**, 33–38.
21. Ormo, M., Cubitt, A. B., Kallio, K., Gross, L. A., Tsien, R. Y., and Remington, S. J. (1996) Crystal structure of the Aequorea victoria green fluorescent protein. *Science* **273**, 1392–1395.

9

Random Oligonucleotide Mutagenesis

Jessica L. Sneeden and Lawrence A. Loeb

1. Introduction

Random oligonucleotide mutagenesis is a method to generate diversity that consists of incorporating random mutations, encoded on a synthetic oligonucleotide, into a specific region of a gene. The number of mutations in individual enzymes within the population can be controlled by varying the length of the target sequence and the degree of randomization during synthesis of the oligonucleotides.

Random oligonucleotide mutagenesis provides a powerful tool for the rapid assembly of large libraries of protein mutants *(1)*. These libraries can be used to explore sequence constraints within a protein and to screen or select for mutants exhibiting phenotypes not found in nature. This approach can be utilized in both prokaryotic and eukaryotic systems. Our lab has used random mutagenesis to study a number of enzymes involved in DNA metabolism. This approach has given us information about the mutability of highly conserved active sites in polymerases *(2–5)* and other enzymes *(6–8)*; in addition it has aided in identification of mutant enzymes that can be used in gene therapy of cancer *(9–12)*. This chapter details construction of a protein library using random oligonucleotide mutagenesis on a plasmid vector for use in positive genetic selection.

There are a number of advantages of this method. First, the mutations introduced are random, but located in a specific region of the gene. This allows targeting of mutagenesis to an area likely to have an effect on desired properties and explores a large amount of sequence space within the given region. While single mutations can give useful information, results can be misleading, as an inactivating mutation does not imply that the amino acid in question is

From: *Methods in Molecular Biology, vol. 231: Directed Evolution Library Creation: Methods and Protocols*
Edited by: F. H. Arnold and G. Georgiou © Humana Press Inc., Totowa, NJ

involved directly in catalysis. In addition, multiple amino acid changes may be required to obtain a desired phenotype.

Another integral feature of the success of random oligonucleotide mutagenesis is its combination with a positive genetic selection. This allows for only those mutants possessing enzyme activity to multiply under certain growth conditions, thereby reducing the number of positives to be analyzed. Iteration further refines the process.

However, some limitations exist in using random oligonucleotide mutagenesis. Because of library size limitations (10^5–10^7), it is usually not possible to explore all possible sequences. For example, given a region of 5 amino acids in length, there are 15 nucleotides. Analysis of all possible sequences at each nucleotide position is 4^{15}, or a total of 1.07×10^9 sequences. This leaves the majority of sequence space unexplored. However, a 100% randomized library will necessarily contain a large fraction of inactive proteins, owing to excessive mutations in a single region and the probability of stop codon incorporation. A decrease of enzyme activity is correlated with an increase in the number of mutations; therefore it is generally advisable to bias library construction such that each individual within the population encodes a more modest number of amino acid changes. This method is region-specific; while exploring much of the sequence space in one specific area, possibly relevant mutations in other parts of the gene are not examined. Combining random oligonucleotide mutagenesis with error-prone PCR mutagenesis or gene shuffling may provide a means of resolving this dilemma.

Random oligonucleotide mutagenesis is illustrated in **Fig. 1**. Oligonucleotides that encode a specific region of the protein of interest are synthesized which contain some determined degree of randomization. The degree of randomization depends on a number of factors, among these are size of the randomized region and its degree of conservation in nature. The result is a library of mutant protein sequences, each of which contains different mutations. These oligonucleotides are then amplified by minimal cycles of PCR to generate double stranded molecules, and cloned by restriction digest into a plasmid-encoded wild-type enzyme, replacing the wild-type region of interest. This creates a library of plasmid-encoded mutant enzymes that can be transformed into an appropriate cell strain and subjected to selection for the desired phenotype. This chapter outlines the steps in creation of a mutant library using random oligonucleotide mutagenesis. Particular aspects of randomization of protein sequence and its constraints are discussed.

2. Materials

1. Suitable vector containing the plasmid-encoded gene of interest.
2. Oligonucleotide containing randomized sequence.

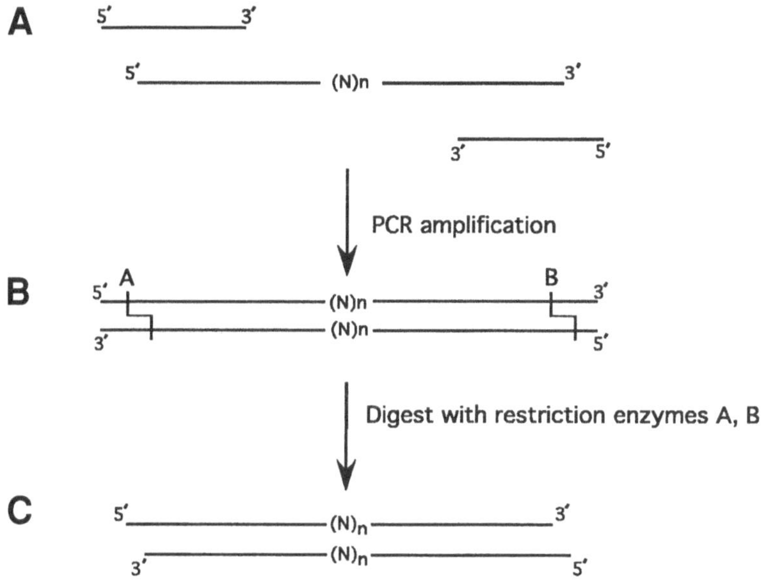

Fig 1. Schematic diagram of oligonucleotide library construction. **(A)** Synthetic oligonucleotide with random sequence (N)n, and flanking primers. Random sequences within an oligonucleotide need not be continuous. **(B)** After PCR amplification, the construct is double-stranded and restriction digested. **(C)** Population of individual random double-stranded oligonucleotides.

3. Primers for amplification of the oligonucleotide.
4. Restriction enzymes, Taq DNA polymerase, and T4 DNA ligase.
5. *E. coli* strain XL-1, DH5α, or other strain suitable for high-efficiency transformation.
6. Luria Bertani Broth (LB) medium.
7. Agarose and DNA sequencing gel equipment.

3. Methods

3.1. Creation of Library Construct

The first step in the generation of a mutant library is the assembly of the oligonucleotide construct. The methods for this are described in **Subheading 3.1.** and include: 1) synthesis of the library oligonucleotide with the desired extent of randomization, and 2) PCR amplification of the oligonucleotide to create a double-stranded molecule for ligation into a plasmid, and restriction digest of the oligonucleotides containing the randomized sequences. The construct generated can then be ligated into a plasmid encoding the gene of interest.

3.1.1. Design of Library Oligonucleotide

Once the region of the enzyme to be randomized has been identified, it is important to identify unique restriction sites adjacent to this region (*see* **Note 1**). If none exist, they can be introduced by silent mutagenesis into the vector *(13)*. The next step is to determine the extent of the randomization of the sequence to be mutated. In general, an increase in the number of amino acid changes correlates with a decrease in protein activity, although there are a number of variables including the number of encoded amino acids and the conserved nature of the region to be randomized (*see* **Note 2**). Oligonucleotides can be synthesized commercially, using standard phosphoramidite methods (*see* **Note 3**).

3.1.2. Amplification and Digestion of Library Sequences

After the synthetic oligo has been generated it can then be amplified using flanking primers (*see* **Note 4** and **Fig. 1A**). Amplification creates a double-stranded oligonucleotide that is necessary for ligation (*see* **Fig. 1B**), but can reduce library diversity through unequal amplification of individuals. Therefore, perform a minimal number of cycles of PCR. PCR is carried out by standard molecular techniques. Briefly, add 2 pmol of the library oligonucleotide template, 20 pmol each flanking primer, 10 m*M* Tris-HCl, pH 9.0 at 25°C, 50 m*M* KCl, 0.1% Triton X-100, 250 μ*M* (total) dNTP mix (dGTP, dCTP, dATP, dTTP), 1 m*M* $MgCl_2$, 20 pmoles each primer, and 2.5U *Taq* polymerase in a total volume of 50 μL H_2O (*see* **Note 5**). Check PCR products by agarose gel electrophoresis *(13)* to ensure correct size and adequate amplification.

The double-stranded oligonucleotide is then digested with the appropriate restriction enzymes, using standard molecular biological methods (*see* **Fig. 1**). For increased ligation efficiency, phenol extract and ethanol precipitate the library oligo, using standard techniques *(13)*.

3.2. Cloning of Library Construct into Plasmid-Encoded Gene

This section outlines the preparation of the vector containing the gene of interest for insertion of the randomized library sequences, insertion of the library oligonucleotide into the vector, and characterization of the resulting library.

3.2.1. Preparation of the Vector

The protein of interest is encoded on a vector suitable for expression. Because of the inefficiency of cloning procedures, it is necessary to remove the wild-type region of interest from the vector prior to ligation of the library sequence, replacing it with a stuffer vector. The stuffer vector should ensure inactivation of the enzyme. Steps for vector preparation are outlined in **Fig. 2**.

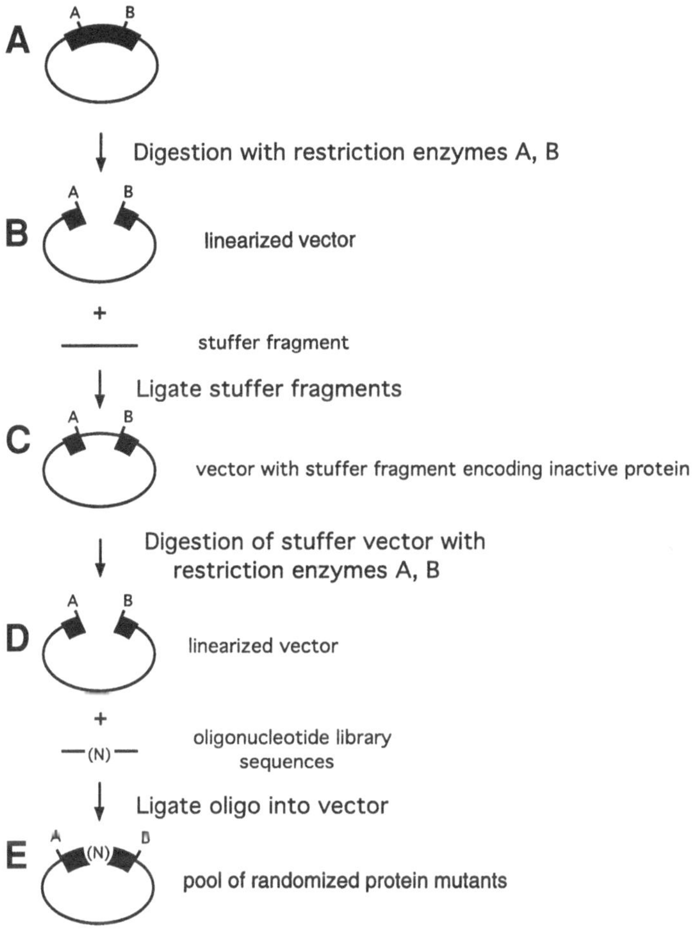

Fig 2. Schematic of insertion of oligonucleotide library into plasmid vector. **(A)** Plasmid-encoded gene with wild-type region of interest is removed by restriction digest at A,B, generating **(B)** linearized vector is ligated to stuffer vector with noncoding sequence, producing **(C)** stuffer vector encoding inactive protein, restricted at sites A and B, to produce **(D)** linearized vector that is ligated to oligonucleotide library, resulting in **(E)** pool of randomized protein mutants.

1. Restriction digest at unique sites A and B (*see* **Fig. 2A**).
2. Ligate a stuffer fragment of DNA into the region of interest (*see* **Note 6**) (*see* **Fig. 2C**).
3. Digest vector containing stuffer fragment with restriction enzymes A and B (*see* **Fig. 2D**).
4. Gel-purify linearized vector (*see* **Note 7**).
5. Ligate oligonucleotide library into purified linearized vector, using standard methods *(13)*.
6. Transform into highly competent cell strain (*see* **Note 8**).

3.2.2. Characterization of the Library

Upon transformation, it is important to determine library size, by determining transformation efficiency. Ideally, a library should be composed of 10^5–10^7 individuals, although useful information may be obtained from somewhat smaller libraries. Library diversity is measured by sequencing at least 50 individual clones that were not yet selected for activity. For this, cells are grown under permissive conditions, or the clones are isolated from a strain that does not require the activity of the enzyme. This pool of unselected sequences can be compared later with selected sequences to determine amino acid residues required for activity (*see* **Note 9**).

3.3. Selection of Proteins with Desired Properties

When selecting for enzyme activity, the use of a positive genetic selection provides a powerful method of library selection. Complementation of a genetic deficiency in a bacterial strain, for example, selects only those mutants possessing catalytic activity. The number of mutants that can be screened is only limited by the transformation efficiency of the cell strain to be used. Ideally, strains that possess a temperature-sensitive defect, or the inability to survive on drug selection should be employed. Where possible, selection conditions that restrict the growth of all inactive proteins should be employed to minimize the number of false postives. Particularly in the selection for enzymes that perform better than wild-type protein, conditions which abolish survival of >99% of cells expressing wild-type should be sought. Libraries can be iteratively selected with increasing levels of stringency, allowing selection for those mutants which possess the desired phenotype as well as a high level of enzyme activity.

4. Notes

1. Note that the further away the restriction sites are from the randomized region, the greater the possibility of protein-inactivating errors due to misalignment during PCR amplification and the possibility of frame-shift error introduction during phosphoramidite synthesis of the library oligonucleotide.
2. One can mutate a region either partially or completely. This will depend on the goal of selection and on the expected tolerance for mutations within the target region. There is a balance between library size and diversity. For example, consider the diversity of a 100% randomized library 10 amino acids in length. Ten amino acids are encoded by 30 nucleotides. For each nucleotide there are four possibilities: G, A, T, and C. Thus, all possible sequences are $4^{30} = 1.15 \times 10^{18}$ individuals. With current cloning and transformation efficiencies, most library sizes are in the range of 10^6–10^8 individuals. Therefore, a library of 10^{18} sequences is not technically achievable and it may be necessary either to randomize this region in smaller separate libraries, or settle for less than complete diversity.

In addition to its diversity limitations, a 100% randomized library will encode more nonfunctional proteins. One way to bias the mutant population toward more enzymatically active proteins is to partially randomize the region of interest. The following calculations are adapted from Suzuki et al. *(7)*. For a given frequency of substitution (p), one can calculate the expected average number of amino acid changes (A) within a given stretch (m) of the region of interest, by using the following equation:

$$A = \sum_{x=1}^{m} a_x$$

The term a_x refers to the sum of the individual amino acid probabilities, and can be calculated for each amino acid in the stretch (m) with the following equations:

one-codon usage (Met, Trp):

$$a_x = 1 - (1 - p)^3 \qquad (1)$$

two-codons usage (Cys, Asp, Glu, Phe, His, Lys, Asn, Cln, Tyr):

$$a_x = 1 - (1 - p)^2(1 - 2p/3) \qquad (2)$$

three-codons usage (Ile): $$a_x = 1 - (1 - p)^2(1 - p/3) \qquad (3)$$

four-codons usage (Ala, Gly, Pro, Thr, Val):

$$a_x = 1 - (1 - p)^2 \qquad (4)$$

six-codons usage (Ser), the value a_x is obtained using either one of three equations, depending on which codons is mutagenized:

for codons UCU/C: $$a_x = 1 - [(1 - p)^2 + (p/3)2(1 - 2p/3)] \qquad (5)$$

for codons UCA/G: $$a_x = 1 - [(1 - p)^2 + (p/3)^2(2p/3)]$$

for codons AGU/C: $$a_x = 1 - [(1 - p)^2(1 - 2p/3) + (p/3)^2]$$

six-codons usage (Leu, Arg), the value a_x is obtained using either one of three equations, depending on which codons is mutagenized:

for codons CGA/G, CUA/G: $$a_x = 1 - [(1 - p)^2 + (1 - p)(p/3)(1 - 2p/3)] \qquad (6)$$

for codons CGU/C, CUU/C: $$a_x = 1 - [(1 - p)^2 + (1 - p)(p/3)(2p/3)]$$

for codons AGA/G, UUA/G: $$a_x = 1 - [(1 - p)^2(1 - 2p/3) + (1 - p)(p/3)]$$

The following is a sample calculation for a library of 5 amino acids randomized to 24% (p = 0.24) at the nucleotide level, for the sequence:

I	S	N	L	K
ATC	AGC	AAC	CTG	AAA

For every nucleotide position, 0.76 is the probability that the nucleotide will be wild-type, and 0.08 is the probability of each of the remaining three nucleotides, for an overall probability (p) of substitution = 0.24.

Using the summation, $A = a_x(I) + a_x(S) + a_x(N) + a_x(L) + a_x(K)$

$$A = [1 - (1 - p)^2(1 - p/3)] + [1 - [(1 - p)^2(1 - 2p/3) + (p/3)^2]] + [1 - (1 - p)^2(1 - 2p/3)] + [1 - [(1 - p)^2 + (1 - p)(p/3)(1 - 2p/3)]] + [1 - (1 - p)^2(1 - 2p/3)]$$

or A = 2.38, the average number of expected amino acid changes per mutant.

For a region of 5 amino acids (15 nucleotides), $0.76^{15} = 0.016$ is the expected frequency of wild-type proteins occurring in the library. This number is important when one expects that the wild-type may be preferentially selected compared to mutants in a given selection.

3. In general, the limit of DNA synthesis by standard phosphoramidite synthesis is about 100 bases. This should be taken into account in library design.
4. Note that restriction digestion of small linear molecules is often inefficient if the restriction site is too close to the end of the oligo. Follow the guidelines for restriction given by the manufacturer for the individual restriction enzymes, and as a general rule design primers with at least 6–10 bases to flank the end of the restriction site.
5. After conditions which amplify the oligo have been optimized, perform 10 reactions to obtain enough material for subsequent cloning.
6. For ease of subsequent purification and to completely inactivate the enzyme, it is advisable to use a large stuffer fragment (around 1 kb) of noncoding sequence. It is important to demonstrate that this "dummy vector" does not possess enzyme activity, before proceeding with library cloning.
7. Gel purification of the linearized vector will decrease the background of inactive dummy vector in the library. There are a number of gel purification protocols ***(13)*** and kits. It should be noted, however, that kits containing NaI should be avoided, as the presence of this salt will decrease ligation efficiencies.
8. Cell strains known to have high transformation efficiencies should be used at this step, rather than direct transformation into the strain to be used for library selection, to maximize library size. Transformation of ligated molecules is generally one to two orders of magnitude lower than supercoiled plasmid ***(13)***. The library can then be isolated and transformed as supercoiled plasmid into other strains for selection.
9. A significant portion of the library will encode inactive protein due to stop codons and frame shifts introduced during synthesis of the library oligonucleotide. These inactivating mutations can constitute up to one-third of the library, which should be taken into account when calculating actual library size.

References

1. Oliphant, A. R. and Struhl, K. (1987) The use of random-sequence oligonucleotides for determining consensus sequences. *Meth. Enzymol.* **155,** 568–582.
2. Glick, E., Vigna, K. L., and Loeb, L. A. (2001) Mutations in human DNA polymerase eta motif II alter bypass of DNA lesions. *EMBO J.* **20,** 7303–7312.
3. Patel, P. H., Kawate, H., Adman, E., Ashbach, M., and Loeb, L. A. (2001) A single highly mutable catalytic site amino acid is critical for DNA polymerase fidelity. *J. Biol. Chem.* **276,** 5044–5051.
4. Patel, P. H. and Loeb L. A. (2000) Multiple amino acid substiutions allow DNA polymerases to synthesize RNA. *J. Biol. Chem.* **275,** 40,266–40,272.
5. Shinkai, A., Patel, P. H., and Loeb, L. A. (2001) The conserved active site motif A of *Escherichia coli* DNA polymerase I is highly mutable. *J. Biol. Chem. 276,* 18,836–18,842.

6. Landis, D. M., Gerlach, J. L., Adman, E. T., and Loeb, L. A. (1999) Tolerance of 5-fluorodeoxyuridine resistant human thymidylate synthases to alterations in active site residues. *Nucleic Acids Res.* **27,** 3702–3711.
7. Suzuki, M., Christians, F. C., Kim, B., Skandalis, A., Black, M. E., and Loeb, L. A. (1996) Tolerance of different proteins for amino acid diversity. *Mol. Diver.* **2,** 111–118.
8. Horwitz, M. S. and Loeb, L. A. (1986). Promoters selected from random DNA sequences. *Proc. Natl. Acad. Sci. USA* **83,** 7405–7409.
9. Encell, L. P. and Loeb, L. A. (1999) Redesigning the substrate specficity of human O^6-alylguanine-DNA alkyltransferase. Mutants with enhanced repair of O^4-methylthymine. *Biochemistry* **38,** 12,097–12,103.
10. Landis, D. M. and Loeb, L. A. (1998) Random sequence mutagenesis and resistance to 5-fluorouridine in human thymidylate synthases. *J. Biol. Chem.* **273,** 25,809–25,817.
11. Christians, F. C., Dawson, B. J., Coates, M. M., and Loeb, L. A. (1997) Creation of human alkyltransferases resistant to O^6-benzylguanine. *Cancer Res.* **57,** 2007–2012.
12. Black, M. E., Newcomb, T. G., Wilson, H-M. P., and Loeb, L. A. (1996) Creation of drug-specific herpes simplex virus type 1 thymidine kinase mutants for gene therapy. *Proc. Natl. Acad. Sci.* **93,** 3525–3529.
13. Sambrook, J. and Russell D. W. (2001) *Molecular Cloning: A Laboratory Manual.* Cold Spring Harbor Laboratory Press, Cold Spring Harbor, NY.

10

Saturation Mutagenesis

Radu Georgescu, Geethani Bandara, and Lianhong Sun

1. Introduction

Directed protein evolution is usually accomplished by generating random or targeted mutations in the protein coding sequence and screening the mutant proteins (library) for functional improvements. By targeting mutations to certain amino acids, we can map the enzyme active site, investigate mechanisms, and study structure-function relationships. With saturation mutagenesis, it is possible to create a library of mutants containing all possible mutations at one or more pre-determined target positions in a gene sequence.

This method is used in directed evolution experiments to expand the number of amino acid substitutions accessible by random mutagenesis *(1)*. In combination with high-throughput screening methods, researchers have successfully used saturation mutagenesis to improve such enzymatic properties as thermostability *(1–3)*, substrate specificity *(4)*, and enantioselectivity *(5)*. It has been shown that a number of homologous enzymes share a common scaffold but catalyze different reactions *(6)*. This implies that an enzyme can change its function by changing only a few residues. Performing combinatorial saturation mutagenesis on a number of such key residues may allow optimization of binding site geometry for new or improved enzymatic properties *(7,8)*.

Saturation mutagenesis can be accomplished by several methods: cassette insertion *(9)*, mutagenic oligonucleotide PCR amplification *(10)*, SOE-ing of DNA fragments *(11)*, or by mutagenic plasmid amplification. One caveat of using mutagenic primers for saturation mutagenesis is that current oligonucleotide synthesis methods have a practical upper limit of ~80 bases. This limits saturation mutagenesis to only a few neighboring codons per primer. It is possible to simultaneously apply saturation mutagenesis to residues that are not nearby in sequence, but the mutation efficiency decreases.

From: *Methods in Molecular Biology, vol. 231: Directed Evolution Library Creation: Methods and Protocols*
Edited by: F. H. Arnold and G. Georgiou © Humana Press Inc., Totowa, NJ

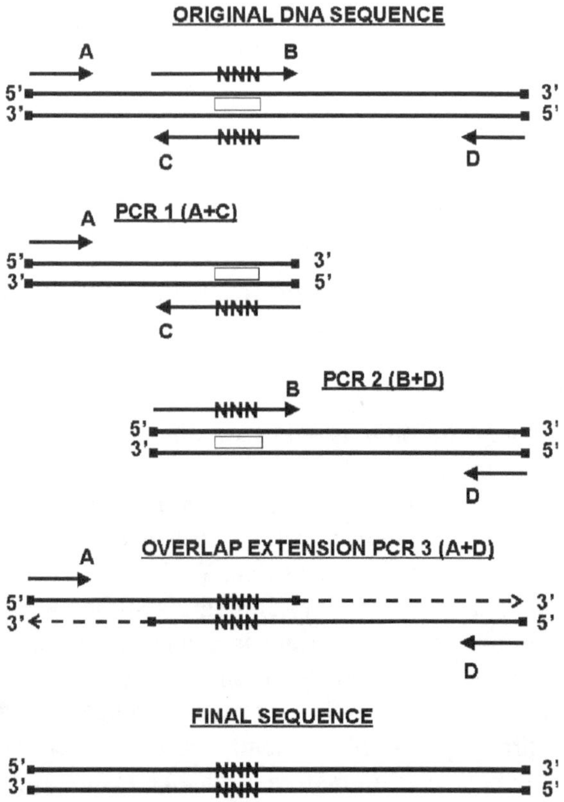

Fig. 1. Saturation mutagenesis by the SOE method. **A** and **D** are the upstream and downstream primers, **B** and **C** are the forward and reverse mutagenic primers, respectively. The codon targeted for mutagenesis is indicated by a rectangle.

We first describe a simple and efficient method based on sequence overlap extension, or SOE *(11)*. In this method, we use separate PCR reactions to amplify two DNA fragments that overlap at specific codon/s in a target sequence. Each primer pair is synthesized with a mismatched random nucleotide sequence in the middle (such as -NNN-), flanked on both sides by nucleotides that specifically anneal to the target region. After an initial PCR amplification, another PCR reaction is used to reassemble a new version of the original full-length sequence (*see* **Fig. 1**), where the target codon/s are effectively randomized.

We also describe a second method based on a simple polymerase-catalyzed replication of a plasmid containing the target sequence. Since lengthy in vitro elongation cycles are subject to replication errors, it is critical to use a high-

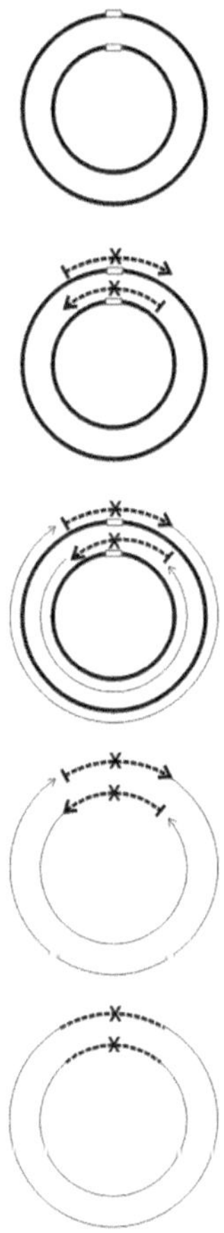

Fig. 2. Saturation mutagenesis by mutagenic plasmid amplification (adapted from Stratagene).

fidelity polymerase such as *Vent* polymerase or *Pfu* polymerase ***(12)***. A pair of mutagenic primers is designed to anneal to the target sequence in a similar fashion to the SOE method described above. After the polymerase initiates elongation at the 3'-end of each primer, replication continues around the entire length of the plasmid (*see* **Fig. 2**). In the final step, the amplified product is digested with *Dpn*I, which recognizes DNA methylated at the sequence 5'-G^{m6}ATC-3'. Since *E. coli* plasmid DNA is methylated while DNA replicated in vitro is not, this step ensures a large excess of replicated vs parental plasmid DNA. No ligation step is necessary here, since the nicks in the replicated DNA are repaired after transformation into *E. coli* cells.

2. Materials

2.1. Biological and Chemical Materials

1. Chemically competent *E. coli* strain(s).
2. Distilled H_2O.
3. Proofreading polymerase (such as *Pfu* polymerase-Stratagene, La Jolla, CA; Vent polymerase- New England Biolabs, Beverly, MA).
4. ZymoClean gel-purification kit (Zymo Research, Orange, CA).
5. Plasmid Mini-prep kit (Qiagen, Chatsworth, CA).
6. Plasmid containing the target sequence to be mutagenized (*see* **Note 1**).
7. Oligonucleotide primers that anneal to upstream and downstream of the target gene.
8. Oligonucleotide primers with the desired mutations (*see* **Note 2** and **Subheading 3.3.**).
9. 10X Reaction buffer: 100 m*M* KCl, 100 m*M* $(NH_4)_2SO_4$, 200 m*M* Tris-HCl, pH 8.8, 20 m*M* $MgSO_4$, 1% Triton X-100.
10. dNTP solution: 10 m*M* of each dNTP (Boehringer-Mannheim, Indianapolis, IN). Store in small aliquots at –20°C.
11. *Dpn*I restriction endonuclease: 20 U/mL (Stratagene, La Jolla, CA).
12. Appropriate restriction enzymes and digestion buffers.
13. T4 DNA Ligase and Ligase Buffer (New England Biolabs, Beverly, MA).

2.2. Equipment

1. Microcentrifuge (Model Eppendorf 5417R, Brinkmann Instruments, Westbury, NY).
2. Thermocycler (Model PTC200, MJ Research, Waltham, MA).
3. Incubator for temperature range 30–62°C (Model Thelco, Precision Scientific, Winchester, VA).
4. Agarose gel electrophoresis supplies and equipment.

3. Methods

3.1. Plasmid Description

To use the saturation mutagenesis protocols, a plasmid containing the cDNA sequence of the gene of interest is needed. For the SOE-ing method, suitable unique flanking restriction sites should be available at both ends of the target DNA sequence.

3.2. Library Size

The size of the saturation mutagenesis libraries is determined by the nature of the genetic code, type of mutagenic codon, and the number of sites that are selected for mutagenesis. Since two amino acids (Trp and Met) are specified by a single codon, a library needs to be large enough to statistically cover all possible substitutions, including these rare amino acids. The library size for one mutated site can be calculated by a binomial probability approximation, where the number of trials equals the library size and the probability of obtain-

Table 1
Calculated Minimum Library Size

Codon type	Possible codons	Library size - 90%	Library size - 95%
NNN	64	144	190
NNG/C or NNG/T	32	72	95

ing either Trp or Met equals the inverse of the total number of possible codons. By specifying a probability of "success" (appearance of Trp or Met) of either 90% or 95%, calculated minimum library sizes are shown in **Table 1**, for a single saturated site.

3.3. Design and Synthesis of Mutagenic Primers

Primers with one or more random codons are used in saturation mutagenesis to combinatorially generate all 20 amino acids, including the wild type. All 64 codons can be used for mutagenesis (by using an NNN random codon), or a limited set can be used (by placing restrictions on the random bases). A common limiting strategy is to restrict the third/wobble position to G or C, by using a NNG/C oligonucleotide. By reducing the number of possible codons to 32, the sequence complexity is decreased and the most common amino acids (Leu, Arg, Ser) are now encoded by only three codons instead of six. In addition, the library sizes are accordingly cut in half and the number of possible "Stop" codons is decreased to one *(13)*.

Each pair of mutagenic primers must be designed taking into account the number and position of the codons chosen for saturation mutagenesis. Both primers contain random codons (NNN or NNG/C) corresponding to the target codons and anneal to the same region on the forward and reverse strands of the target sequence. In addition, the random codons should be flanked by ~10–15 correct bases on each side to allow specific annealing. The GC content needs to be a minimum of ~40% and the primers should terminate in one or more G/C bases. It is important to purify the primers after synthesis by liquid chromatography (FPLC or HPLC) or by polyacrylamide gel electrophoresis (PAGE).

3.4. Library Construction by SOE

This method uses PCR to amplify overlapping pieces of the target gene, using a combination of mutagenic and upstream/downstream primers. These pieces are subsequently used to assemble a new version of the parent gene.

1. Using thin-walled PCR tubes, set up two separate reactions (PCR 1 and PCR 2) containing: 0.1–100 ng of template DNA, 1X polymerase buffer, 250 μ*M* dNTP mix, 100–150 ng of each primer (*see* **Fig. 1** for primer combinations), 2.5 U thermostable proofreading DNA polymerase, and sterile distilled water to 50 μL.

2. Run both PCR reactions with the following cycling protocol: 95°C for 2 min, 20–30 cycles of (95°C for 30 s, 52°C for 30 s, 72°C for 2 min), 72°C for 10 min (*see* **Note 1**).
3. PCR 3: using a thin-walled PCR tube, mix 7–10 µL DNA samples each from PCR 1 and PCR 2, 100–150 ng of upstream and downstream primers, 1X polymerase buffer, 250 µ*M* dNTP mix, 2.5 U thermostable proofreading DNA polymerase and sterile distilled water to 50 µL.
4. Run the PCR 3 reaction with the protocol from **Step 2**. After completion, run a small aliquot (1–4 µL) of the PCR-amplified DNA on an agarose gel to check product yield and purity.
5. From the observed migration distance, calculate the approximate size of each product and match it with the expected fragment size calculated from the sequence/primer information. After PCR 3, the size of the assembled DNA should closely match the size of the original sequence.
6. Gel-purify the PCR 3 reaction products using the Zymoclean Gel DNA Recovery Kit (follow the manufacturer's protocol).
7. Digest the reaction product and the original plasmid with the appropriate restriction enzymes. This creates compatible sticky ends that allow the PCR products to be ligated back into the initial vector.
8. Again, run the digested products on an agarose gel to check size and completion of digestion. Gel-purify the digested reaction products with the Zymoclean Gel DNA Recovery Kit.
9. Ligate the compatible sticky-end fragments (empty vector and insert) overnight at 16°C by using the manufacturer's ligation protocol for T4 DNA Ligase (*see* **Notes 2–3**).
10. The DNA library is now ready for transformation.

3.5. Library Construction by Plasmid Amplification

This method uses PCR with a proofreading enzyme and mutagenic primers to amplify the plasmid containing the target sequence (*see* **Note 4**).

1. In a thin-walled PCR tube, combine 5–50 ng of template DNA, 125 ng of each primer, 1X PCR buffer, 200 µ*M* dNTP mix, 2.5 U polymerase, and sterile distilled water to 50 µL.
2. Run a PCR reaction using the following cycling protocol: initial denaturation at 95°C for 30 s followed by 16 cycles of denaturation at 95°C for 30 s, annealing at 55°C for 1 min, and extension at 68°C for 2 min for each kb of plasmid length.
3. After completion, run 10 µL aliquots of the reactions on a 1% agarose gel to check for sufficient amplification. Select the reactions with high yield.
4. Add 1 µL of *Dpn*I enzyme (10 U/µL) directly to each amplification mixture, mix gently and incubate for 1 h at 37°C.
5. The DNA library is now ready for transformation (*see* **Note 5**).

3.6. Multiple-Site Saturation Mutagenesis

As an extension to the methods previously described, one can also construct multiple-site saturation mutagenesis libraries. For sites close by in sequence, multiple-site saturation can be achieved with SOE or plasmid amplification, by using long mutagenic primers containing NNN or NNG/C sequences at every mutated site. If the sites targeted for saturation are distant in sequence, an alternative protocol can be used:

3.6.1. Construction of Multiple-Site Saturation Libraries Using SOE

1. Amplified DNA fragments containing single-site saturation mutation can be used as templates to create a double-site saturation mutagenesis library.
2. Repeat this process to introduce more sites (*see* **Note 6**).

3.6.2. Construction of Multiple-Site Saturation Libraries Using Plasmid Amplification

1. Start an *E. coli* culture by inoculating growth medium with more than 200 single colonies from a plate containing single-site saturation mutants, prepared by plasmid amplification.
2. Purify the plasmid DNA from the cell culture and use it as a template for a second SOE or plasmid amplification. This creates a double-site saturation mutagenesis library.
3. Go back to the first step and repeat the cycle for each additional site (*see* **Note 6**).

4. Notes

1. If the final step in overlap extension PCR does not give any products, check the length of actual overlap between the gene fragments. If the overlap is not good enough to cause inter-strand annealing (not enough GC content or too few matching base pairs), no product will be generated from the PCR reaction. Possible solutions are to increase the primer length to increase the overlap or to drop the PCR annealing temperature by 5–10°C to promote annealing. SOE mutagenesis can be tricky when a target site is close to either end of the gene, since one of the resulting PCR fragments will be small and difficult to isolate. A useful workaround is to increase the fragment size by using a redesigned PCR primer that anneals outside the respective end of the gene but remains within the boundaries of the restriction site.
2. Ligation efficiency can be improved by decreasing the total mixture volume to 10 μL, doubling the ligase amount, using super-concentrated T4 ligase enzyme, or increasing incubation time from a few hours to overnight.
3. For best ligation results, set up several experimental conditions where the quantity of inserted DNA is in excess over the cut vector DNA. Molar ratios of 3:1 and 5:1 (as estimated from OD_{260} values) are good starting points.
4. To minimize parental DNA that may contribute to non-mutated clones, the plasmids used for templates should be isolated from a methylation positive *dam*$^+$

E. coli strain, e.g., XL10-Gold (Stratagene, La Jolla, CA). This allows digestion of the parental DNA by *Dpn*I.

5. High-transformation efficiency *E. coli* strains like XL-10 Gold (Stratagene, La Jollla, CA) or Z competent *E. coli* strains (Zymo Research, Orange, CA) should be used according to manufacturer's instructions. Purification of *Dpn*I-treated plasmid using a PCR purification kit can greatly increase transformation efficiency.
6. As the number of saturated sites increases, multiple-site saturation libraries need to cover all combinatorial possibilities. Considering also the decreasing experimental efficiency for each additional cycle that generates multiple mutants, it appears that multiple-site saturation mutagenesis is practically limited to a small number of sites.

References

1. Miyazaki, K. and Arnold, F. H. (1999) Exploring nonnatural evolutionary pathways by saturation mutagenesis: rapid improvement of protein function. *J. Mol. Evol.* **49,** 716–720.
2. Miyazaki, K., Wintrode, P. L., Grayling, R. A., Rubingh, D. N., and Arnold, F. H. (2000) Directed evolution study of temperature adaptation in a psychrophilic enzyme. *J. Mol. Biol.* **297,** 1015–1026.
3. Morawski, B., Quan, S., and Arnold, F. H. (2001) Functional expression and stabilization of horseradish peroxidase by directed evolution in Saccharomyces cerevisiae. *Biotechnol. Bioeng.* **76,** 99–107.
4. Sakamoto, T., Joern, J. M., Arisawa, A., and Arnold, F. H. (2001) Laboratory evolution of toluene dioxygenase to accept 4-picoline as a substrate. *Appl. Environ. Microbiol.* **67,** 3882–3887.
5. May, O., Nguyen, P. T., and Arnold, F. H. (2000) Inverting enantioselectivity by directed evolution of hydantoinase for improved production of L-methionine. *Nat. Biotechnol.* **18,** 317–320.
6. Babbitt, P. C. and Gerlt, J. A. (1997) Understanding enzyme superfamilies. Chemistry As the fundamental determinant in the evolution of new catalytic activities. *J. Biol. Chem.* **272,** 30,591–30,594.
7. Horwitz, M. S. and Loeb, L. A. (1986) Promoters selected from random DNA sequences. *Proc. Natl. Acad. Sci. USA* **83,** 7405–7409.
8. Whittle, E. and Shanklin, J. (2001) Engineering delta 9-16:0-acyl carrier protein (ACP) desaturase specificity based on combinatorial saturation mutagenesis and logical redesign of the castor delta 9-18:0-ACP desaturase. *J. Biol. Chem.* **276,** 21,500–21,505.
9. Kegler-Ebo, D., Pollack, G., and DiMaio, D. (1996) Use of Codon Cassette Mutagenesis for Saturation Mutagenesis, in *In Vitro Mutagenesis Protocols* (Trower, M. K., ed.), Humana Press, Totowa, NJ, pp. 297–311.
10. Chiang, L. W. (1996) Saturation Mutagenesis by Mutagenic Oligonucleotide-Directed PCR Amplification (Mod-PCR), in *In Vitro Mutagenesis Protocols* (Trower, M. K., ed.), Humana Press, Totowa, NJ, pp. 311–321.

11. Ho, S. N., Hunt, H. D., Horton, R. M., Pullen, J. K., and Pease, L. R. (1989) Site-directed mutagenesis by overlap extension using the polymerase chain reaction. *Gene* **77,** 51–59.
12. Cline, J., Braman, J. C., and Hogrefe, H. H. (1996) PCR fidelity of *Pfu* DNA polymerase and other thermostable DNA polymerases. *Nucleic Acids Res.* **24,** 3546–3551.
13. Reidhaar-Olson, J. F., Bowie, J. U., Breyer, R. M., et al. (1991) Random mutagenesis of protein sequences using oligonucleotide cassettes. *Meth. Enzymol.* **208,** 564–586.

11

DNA Shuffling

John M. Joern

1. Introduction

DNA shuffling is a method for in vitro recombination of homologous genes invented by W.P.C Stemmer *(1)*. The genes to be recombined are randomly fragmented by DNaseI, and fragments of the desired size are purified from an agarose gel. These fragments are then reassembled using cycles of denaturation, annealing, and extension by a polymerase (*see* **Fig. 1**). Recombination occurs when fragments from different parents anneal at a region of high sequence identity. Following this reassembly reaction, PCR amplification with primers is used to generate full-length chimeras suitable for cloning into an expression vector.

In several instances, chimeric enzymes with improved activity and stability have been isolated from libraries constructed using DNA shuffling *(2–5)*. In other cases, the method resulted in libraries with either too many mutations *(6)* or too few crossovers *(7)* to be useful. The DNA shuffling method we describe in this chapter is a hybrid of various published methods that has yielded highly chimeric libraries (as many as 3.7 crossovers per 2.1 kb gene) with a low mutagenesis rate *(8)*. Fragments are made in much the same way as in the first Stemmer method *(1)*, the reassembly protocol is borrowed from Abècassis, et al. *(9)*, and *Pfu* polymerase is used throughout, as suggested by Zhao, et al. *(6)*. We have used this method successfully to recombine parents with only 63% DNA sequence identity; however, more crossovers occur (and the library is more diverse) when the parent genes are more similar *(8)*.

2. Materials

1. Cloned *Pfu* polymerase and 10X buffer (Stratagene, La Jolla, CA).
2. PCR nucleotide mix, 10 m*M* each (Promega, Madison, WI).
3. Dimethyl sulfoxide (DMSO).

From: *Methods in Molecular Biology, vol. 231: Directed Evolution Library Creation: Methods and Protocols*
Edited by: F. H. Arnold and G. Georgiou © Humana Press Inc., Totowa, NJ

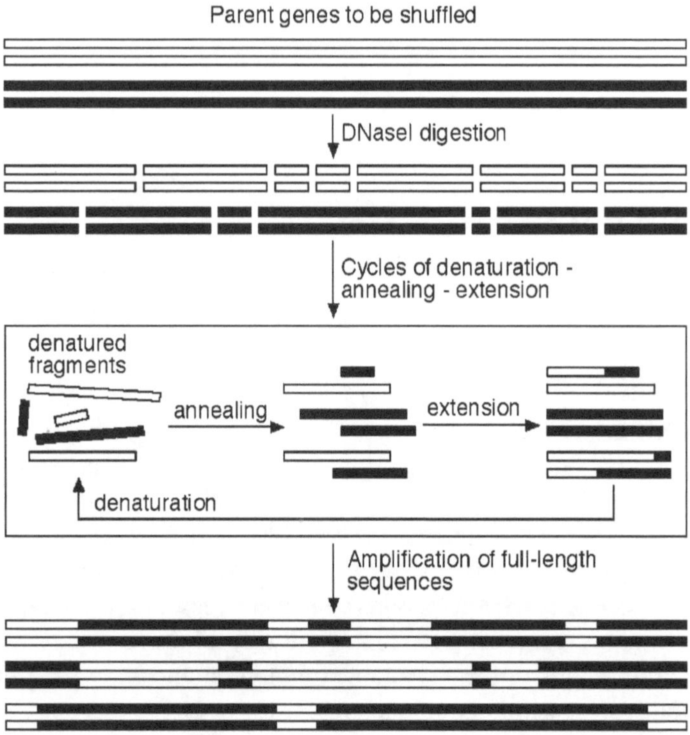

Fig. 1. Schematic of DNA shuffling method. Parental genes are cleaved randomly using DNaseI to generate a pool of fragments. These fragments are recombined using PCR with a specialized thermocycling protocol. Fragments are denatured at high temperature, then allowed to anneal to other fragments. Some of these annealing events result in heteroduplexes of fragments from two homologous parents. Annealed 3' ends are then extended by polymerase. After 20–50 cycles of assembly, a PCR amplification with primers is used to selectively amplify full-length sequences.

4. MJ Research PTC-200 thermal cycler.
5. 0.5 *M* Tris-HCl, pH 7.4.
6. 0.5 *M* ethylenediaminetetraacetic acid (EDTA), pH 8.0.
7. 0.2 *M* manganese chloride.
8. DNaseI, Type II, from bovine pancreas (Sigma, St. Louis, MO).
9. QIAquick gel extraction kit (QIAGEN: Valencia, CA) or equivalent.
10. G25 Columns (Amersham Pharmacia Biotech, Inc., Piscataway, NJ).
11. Two sets of primers (*see* **Note 1**).
12. Parent DNA. The parent DNA should have large regions flanking the gene of interest so that "nested" primers can be used (*see* **Note 1**). A plasmid containing the gene of interest is ideal.

3. Methods

The procedures outlined below detail 1) obtaining DNA fragments from a DNaseI digestion, 2) reassembly of those fragments, and 3) amplification of full-length sequences from the reassembly reaction.

3.1. Obtaining DNA Fragments for Shuffling

1. To get parent DNA for shuffling, mix in a PCR tube: 10 μL 10X *Pfu* buffer, 2 μL of PCR nucleotide mix, 40 pmol of each outer primer (*see* **Note 1**), 5 U of *Pfu* polymerase, 3 μL DMSO, 0.08 pmol of template, and 75 μL of water. Thermocycle using an annealing temperature appropriate for the outer primers. Extension should occur at 72°C for 2–3 min per kilobase of DNA amplified. 20–25 cycles are generally required.
2. Using a QIAquick gel extraction kit or similar spin column system, purify the PCR reactions. The DNA concentration after purification should be at least 40 μg/mL.
3. For the DNaseI fragmentation, prepare a solution of 0.167 *M* Tris-HCl buffer, 0.0833 M manganese chloride, and 1.67 U/mL DNaseI (*see* **Note 2**). In a separate tube, prepare 70 μL of an equimolar mix of parent DNA with a concentration of 50–125 μg/mL. Bring these solutions to 15°C in a thermocycler. At the same time, put 6 μL of EDTA solution on ice in a microcentrifuge tube. Add 30 μL of the buffered DNaseI solution into the parent DNA mix, and mix by pipetting several times. Incubate at 15°C for 0.5–10 min (*see* **Note 3**). To stop the reaction, transfer the solution to the tube containing EDTA and mix thoroughly.
4. Run the DNA fragments on an agarose gel containing ethidium bromide, and excise the desired size range (*see* **Note 4**). Purify the selected fragments using a QIAquick gel extraction kit or similar spin column system. The effluent should be further purified using a G25 column.

3.2. Reassembly of DNaseI Fragments

1. To 42 μL of purified fragment DNA, add 5 μL of 10X *Pfu* buffer, 2 μL of dNTP solution, and 1 μL of *Pfu*.
2. Cycle according to the following protocol: 96°C, 90 s; 35 cycles of (94°C, 30 s; 65°C, 90 s; 62°C, 90 s; 59°C, 90 s; 56°C, 90 s; 53°C, 90 s; 50°C, 90 s; 47°C, 90 s; 44°C, 90 s; 41°C, 90 s; 72°C, 4 min); 72°C, 7 min; 4°C thereafter (*see* **Note 5**).
3. Run 5 μL of this reaction on an agarose/ethidium bromide gel. A smear of reassembled DNA that extends above the molecular weight of the parent genes should be visible.

3.3. Amplification of Full-Length Sequences

1. Combine 10 μL of 10X *Pfu* buffer, 2 μL of dNTP solution, 40 pmol of each inner primer, 3 μL of DMSO, an aliquot (10–1000 nL) of unpurified reassembly reaction (*see* **Note 6**), 5 U of *Pfu*, and water to a final volume of 100 μL.
2. Thermocycle using an annealing temperature appropriate for the inner primers. Extension should occur at 72°C for 2–3 min per kilobase of DNA amplified. 20–25 cycles are generally required.

3. Run 5 μL of this reaction on an agarose/ethidium bromide gel. A band should be observed at the molecular weight of the parent gene. DNA should be purified by gel extraction prior to cloning into an expression vector.

4. Notes

1. Design of primer sets. Two sets of 18–25 bp primers with GC content ~50% should be designed in a "nested" configuration, i.e., the inner primers close to the gene of interest, and the outer primers ~150 bp outside of the inner primers. The outer primers are used to amplify DNA for the fragmentation reaction, and the inner primers are used to amplify full-length sequences following the assembly reaction. Generally, when only one primer set is used, the amplification step to regenerate full-length sequences will fail. This might result from digestion or degradation of priming sites during the reassembly due to residual exonuclease activity from the polymerase.
2. Handling of DNaseI. DNaseI was dissolved in sterile water to a concentration of 10 U/μL and stored at –20°C. An aliquot from a fresh 1:200 dilution was used to carry out the DNaseI digestion protocol.
3. Incubation with DNaseI. The incubation time with DNaseI is a critical parameter for generating fragments of the desired size. Before attempting this step, prepare enough parent DNA to digest small aliquots with varying incubation times (30 s to 10 min). Then, select an optimal condition for a larger-scale digestion. In our hands, digestion of a 2.6 kb gene for 2 min gave a size distribution of fragments centered at ~0.7 kb.
4. Selecting an appropriate fragment size. The first account of DNA shuffling reported selecting fragments in the range of 10–50 bp *(1)*. This size range can be difficult to reassemble. Using this method, we have had success using fragments of 0.4–1 kb to reassemble a 2.1 kb gene and create a chimeric library with 3.7 crossovers per gene *(8)*. Thus, if 10–50 bp fragments do not reassemble successfully, using larger fragments may get the reassembly to go while still generating a sufficiently diverse library.
5. Temperature cycle during reassembly. Alternatively, a constant annealing temperature can be used for the reassembly. We had success annealing at constant temperatures ranging from 42°C to 58°C for 5 min. For small fragments (~100–500 bp) a higher annealing temperature (58°C) was required to eventually obtain a full-length product (2.1 kb), but a set of large fragments (~200–1500 bp) reassembled readily using either a 42, 50, or 58°C annealing temperature (unpublished results). Theoretically, more crossovers should occur when a lower annealing temperature is used, and we have in fact observed this experimentally (unpublished results).
6. Amplification of full-length sequences from reassembly reaction. In our experience, this step requires the most optimization. The amount of assembly reaction and the number of cycles are critical variables. We suggest varying the number of cycles from 20 to 25 cycles, and the amount of reassembly reaction from 1 μL to 10 nL per 100 μL reaction. We have observed the counterintuitive result that if

too many cycles (28–32) are used, a significant decrease in yield occurs. If too much reassembly reaction was added, a smear was observed upon running the reaction on a gel.

7. Mutagenesis rate. Using this method, we shuffled three parent genes of 2.1 kb and sequenced 8 active chimeras and 10 inactive chimeras. Only four spontaneously generated mutations were found for a nucleotide mutation rate of 0.011%. If mutations are desired in addition to recombination, error-prone PCR can be used in the first step to amplify parent DNA for the DNaseI digestion.

References

1. Stemmer, W. P. C. (1994) DNA shuffling by random fragmentation and reassembly: In vitro recombination for molecular evolution. *Proc. Natl. Acad. Sci. USA* **91,** 10,747–10,751.
2. Chang, C. J., Chen, T. T., Cox, B. W., et al. (1999) Evolution of a cytokine using DNA family shuffling. *Nat. Biotechnol.* **17,** 793–797.
3. Ness, J. E., Welch, M., Giver, L., et al. (1999) DNA shuffling of subgenomic sequences of subtilisin. *Nat. Biotechnol.* **17,** 893–896.
4. Christians, F. C., Scapozza, L., Crameri, A., Folkers, G. and Stemmer, W. P. C. (1999) Directed evolution of thymidine kinase for AZT phosphorylation using DNA family shuffling. *Nat. Biotechnol.* **17,** 259–264.
5. Bruhlmann, F. and Chen, W. (1998) Tuning biphenyl dioxygenase for extended substrate specificity. *Biotech. Bioeng.* **63,** 544–551.
6. Zhao, H. and Arnold, F. H. (1997) Optimization of DNA shuffling for high fidelity recombination. *Nucleic Acids Res.* **25,** 1307–1308.
7. Kikuchi, M., Ohnishi, K., and Harayama, S. (1999) Novel family shuffling methods for the in vitro evolution of enzymes. *Gene* **236,** 159–167.
8. Joern, J. M., Meinhold, P., and Arnold, F. H. (2002) Analysis of Shuffled Gene Libraries. *J. Mol. Biol.* **316,** 643–656.
9. Abècassis, V., Pompon, D., and Truan, G. (2000) High efficiency family shuffling based on multi-step PCR and in vivo DNA recombination in yeast: statistical and functional analysis of a combinatorial library between human cytochrome P450 1A1 and 1A2. *Nucleic Acids Res.* **28,** e88.

12

Family Shuffling with Single-Stranded DNA

Wenjuan Zha, Tongbo Zhu, and Huimin Zhao

1. Introduction

Family shuffling, which generates chimeric progeny genes by recombining a set of naturally occurring homologous genes, is an extremely powerful approach for in vitro protein evolution. In comparison with other in vitro protein evolution methods, family shuffling has the advantage of sampling a larger portion of the sequence space that has been proven functionally rich by nature. So far, the most widely used technique to carry out family shuffling is DNA shuffling *(1–5)*. However, one significant drawback associated with this method is the low frequency of chimeric genes (recombined gene products) in the shuffled library *(6)*, which may be largely owing to the annealing of DNA fragments derived from the same parental genes (homo-duplex formation) whose probability is much higher than that of hetero-duplex formation. This is true even when the sequence homologies among parental genes are higher than 80% *(7)*.

To address this problem, Kikuchi et al. *(7)* developed a modified family shuffling method—family shuffling with restriction enzyme-cleaved DNA fragments—based on the DNA shuffling method, which involves the fragmentation of the parental genes using restriction enzymes rather than using DNase I. In addition, the same research laboratory developed a second modified family shuffling method, family shuffling with single-stranded DNA (ssDNA) *(8)*. The latter method is also based on the DNA shuffling method and uses single-stranded DNA templates rather than double-stranded DNA (dsDNA) templates for DNase I fragmentation. Since use of single-stranded DNA as templates will decrease the probability of homo-duplex formation, the percentage of the parental genes in the shuffled library should be significantly reduced. As a demonstration, this method was used to recombine two catechol 2,3-

From: *Methods in Molecular Biology, vol. 231: Directed Evolution Library Creation: Methods and Protocols*
Edited by: F. H. Arnold and G. Georgiou © Humana Press Inc., Totowa, NJ

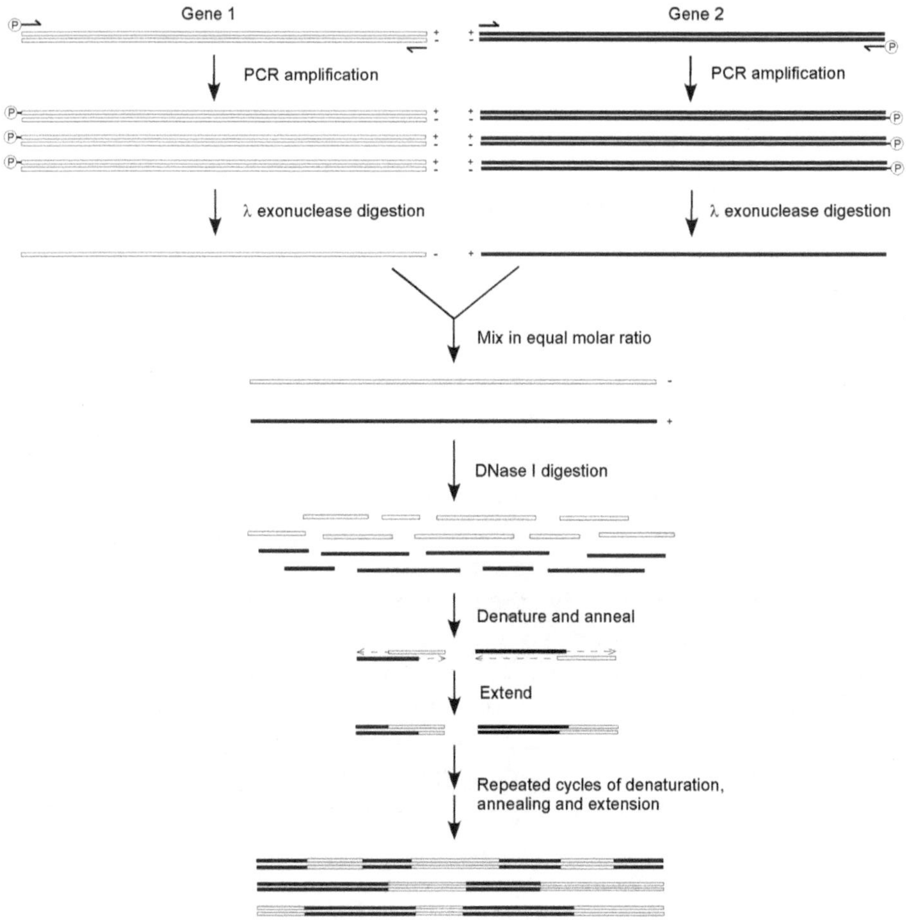

Fig. 1. Schematic representation of ssDNA-based family shuffling. For simplicity, only two DNA templates are shown. Double stranded DNAs with one strand phosphorylated at 5'-terminus are prepared from DNA templates carrying the target sequences using conventional PCR and digested by lambda exonuclease. The two single-stranded DNA templates are mixed at equal molar ratio followed by random fragmentation with DNase I. After the removal of the oligonucleotides and the templates, the homologous fragments are reassembled into full-length chimeric genes in a PCR-like process. The full-length genes may be amplified by a standard PCR and sub-cloned into an appropriate vector.

dioxygenase genes, *nahH* and *xyl*E. It was found that this ssDNA-based DNA shuffling was able to generate chimeric genes at a rate of 14%, much higher than the rate of less than 1% obtained by the original dsDNA-based DNA shuffling method *(8)*. However, the method Kikuchi et al. used to prepare single-stranded DNA templates requires the subcloning of the target genes to a phagemid vector and the use of a helper phage. The procedure is time-consuming and inefficient. Moreover, the isolated single-stranded DNA templates all contain the backbone DNA of the phagemid vector, which could become problematic in certain cases since the DNA fragments from the backbone DNA may interfere with the fragment reassembly process of the parental genes.

Here we describe an improved ssDNA-based family shuffling method in which a simple and convenient approach is used to prepare single-stranded DNA templates. The procedure is illustrated in **Fig. 1**. For simplicity, only two parental genes are shown. First, double-stranded DNAs with one strand phosphorylated at 5'-terminus are prepared from DNA templates carrying the target sequences using conventional PCR. For each target sequence, only one of the two PCR primers is 5'-phosphorylated. The primers should be designed such that the 5'-phosphorylated strands from different parental genes are complementary to each other. Second, single-stranded DNA templates are prepared using lambda exonuclease digestion. In the presence of lambda exonuclease, the 5'-phosphorylated strands will be degraded very rapidly. Third, the two single-stranded DNA templates are mixed at equal molar ratio, followed by random fragmentation with DNase I. Fourth, the digested single-stranded DNA fragments are reassembled into full-length chimeric genes.

As a test case, we used this method to shuffle two genes encoding the ligand binding domain of estrogen receptor α and β (ERα-LBD and ERβ-LBD), which share 66% sequence homology at the DNA level. The shuffled gene products were subcloned into an *E. coli* vector pET26b(+) (Novagen), and five transformants were randomly selected for DNA sequencing. It was found that all five genes were chimeric. In comparison, we also shuffled ERα-LBD and ERβ-LBD using the conventional dsDNA based DNA shuffling method. None of the five randomly selected transformants obtained from the shuffled products were chimeric genes. Therefore, it appears that ssDNA based DNA shuffling is indeed a more efficient family shuffling method than dsDNA based DNA shuffling, especially when the parental genes to be shuffled have relatively low sequence homology.

2. Materials

1. DNA templates containing the target sequences to be recombined (*see* **Note 1**).
2. Oligonucleotide primers (*see* **Note 2**).
3. Lamda exonuclease and its 10X reaction buffer (New England BioLabs, Beverly, MA).
4. 10X DNase I digestion buffer: 500 m*M* Tris-HCl, pH 7.4, 100 m*M* $MnCl_2$.
5. 10 m*M* Tris-HCl, pH 7.4.
6. DNase I: 10 U/μL (Roche Diagnostics, Indianapolis, IN).
7. *Taq* DNA polymerase and its 10X reaction buffer (Promega, Madison, WI).

8. *Pfu*Turbo DNA polymerase and its 10X reaction buffer (Stratagene, La Jolla, CA).
9. 10X dNTP mix: 2 m*M* of each dNTP (Roche Diagnostics, Indianapolis, IN).
10. Agarose gel electrophoresis supplies and equipment.
11. MJ PTC-200 thermocycler (MJ Research Inc., Watertown, MA).
12. EZ load precision molecular mass ruler (Bio-Rad, Hercules, CA).
13. QIAquick PCR purification kit (QIAgen, Valencia, CA).
14. QIAEX II gel extraction kit (QIAgen, Valencia, CA).
15. Centri-Sep columns (Princeton Separations, Inc., Adelphia, NJ).

3. Methods

1. For each target sequence, combine 1–10 ng DNA template, 10X *Taq* reaction buffer, 0.5 μ*M* each primer (one of the two primers is 5'-phosphorylated), 10X dNTP mix (2 m*M* each dNTP), and 2.5 U *Pfu/Taq* DNA polymerase (1:1) in a total volume of 100 μL (*see* **Note 3**).
2. Run the PCR reaction using the following program: 96°C for 2 min, 30 cycles of 94°C for 30 s, 55°C for 30 s, and 72°C for 1 min, followed by 72°C for 7 min (*see* **Note 4**).
3. Purify the PCR products according to the manufacturer's protocol in the QIAquick PCR purification kit. Estimate DNA concentrations using gel electrophoresis and EZ load precision molecular mass ruler or using a UV-Vis spectrophotometer.
4. For each target gene, combine 3–5 μg PCR product, 10X lambda exonuclease reaction buffer, and 10 U lambda exonuclease in a total volume of 50 μL. Incubate the reaction mixture at 37°C for 30 min–60 min (*see* **Note 5**).
5. Separate the digestion products by gel electrophoresis. Purify the single-stranded DNA products using a QIAEX II gel purification kit (*see* **Note 6**). Estimate DNA concentrations using gel electrophoresis and EZ load precision molecular mass ruler (*see* **Note 7**).
6. Mix ~0.5 μg of each purified ssDNA product. Dilute the mixture to 45 μL in 10 m*M* Tris-HCl, pH 7.4, and add 5 μL of 10X DNase I digestion buffer.
7. Equilibrate the mixture at 15°C for 5 min before adding 0.6 U of DNase I. Digestion is done on a thermocycler using the following program: 15°C for 2 min and 90°C for 10 min (*see* **Note 8**).
8. Purify the cleaved DNA fragments using Centri-Sep columns according to the manufacturer's protocol.
9. Combine 10 μL of the purified fragments, 2 μL of 10X *Taq* buffer, 2 μL of 10X dNTP mix, 0.5 U *Taq* DNA polymerase and sterile dH_2O in a total volume of 20 μL (*see* **Note 9**). No primer is added.
10. Run the reassembly reaction using the following program: 3 min at 96°C followed by 40 cycles of 30 s at 94°C, 1 min at 55°C, 1 min + 5 s/cycle at 72°C, and finally 7 min at 72°C (*see* **Note 10**).
11. Run a small aliquot of the reaction mixture on an agarose gel. A smear extending through the size of the expected full-length gene should be visible. Otherwise, run an additional 10–20 cycles, or lower the annealing temperature.

12. Combine 1 μL of the assembly reaction, 0.3–1.0 μ*M* each primer, 10 μL of 10X *Taq* buffer, 10 μL of 10X dNTP mix (2 m*M* each dNTP), and 2.5 U *Taq/Pfu* (1:1) mixture in a total volume of 100 μL (*see* **Note 11**).
13. Run the PCR reaction using the following program: 96°C for 2 min, 10 cycles of denaturation at 94°C for 30 s, annealing at 55°C for 30 s and elongation at 72°C for 45 s followed by another 14 cycles of 30 s at 94°C, 30 s at 55 °C and 45 s + 20 s/cycle at 72°C. The final step of elongation is at 72°C for 7 min.
14. Run a small aliquot of the reaction mixture on an agarose gel. In most cases, there is only a single band at the correct size. Purify the full-length PCR product using the QIAquick PCR purification kit, or other reliable method. If there is more than one band, purify the product of correct size using the QIAEX II gel purification kit.
15. Digest the product with the appropriate restriction endonucleases, and ligate to the desired cloning vector.

4. Notes

1. Appropriate templates include plasmids carrying target sequences, cDNA or genomic DNA carrying the target sequences, sequences excised by restriction endonucleases, and PCR amplified sequences.
2. Primer design should follow standard criteria including similar melting temperatures and elimination of self-complementarity or complementarity of primers to each other. Free computer programs, such as Primer3 at Biology Workbench (http://workbench.sdsc.edu), can be used to design primers. Typically, primers should also include unique restriction sites for subsequent directional sub-cloning. For each target sequence, only one of the two primers used to amplify the DNA should be 5'-phosphorylated. In the case of two target sequences, primers should be 5'-phosphorylated in a complementary manner (*see* **Fig. 1**), such that the resultant single-stranded DNAs are complementary. In the case of more than two target sequences, one of the target sequences is selected to generate a single-stranded DNA complementary to those from the rest of target sequences. 5'-Phosphorylated primers can either be ordered directly from various DNA synthesis service providers or prepared from regular primers using T4 polynucleotide kinase.
3. *Pfu*Turbo DNA polymerase is selected because of its high fidelity in DNA amplification and the generation of blunt-ended products. Other DNA polymerases with proofreading activity such as *Vent* DNA polymerase (New England Biolabs, Beverly, MA) or a mixture of *Taq* DNA polymerase and *Pfu* DNA polymerase (1:1) may also be used. *Taq* DNA polymerase alone is not recommended since it cannot generate blunt-ended and 5'-phosphorylated products that are preferred substrates for lambda exonuclease.
4. The elongation time depends on the gene size. For genes of 1 kb or less, an elongation time of 1 min will suffice. For genes larger than 1 kb, the elongation time should be adjusted accordingly. If the yield of the PCR product is low, set up several PCR reactions for each target sequence and combine the products.
5. The extent of the digestion reaction should be checked by running a 2 μL aliquot of the reaction mixture along with an aliquot of undigested DNA as a control on

an agarose gel. Single-stranded DNA will migrate faster than the double-stranded counterpart. Note that the extent of phosphorylation of the PCR primers is critical for lambda exonuclease digestion. It is likely that some double-stranded DNA will still be present in the reaction mixture even after digestion for a long time or with excess enzymes. Those undigested DNAs are presumably the PCR products derived from the non-phosphorylated primers. Furthermore, since lambda exonuclease will also degrade single-stranded DNA substrates, albeit at a greatly reduced rate, digestion of the PCR products for a long time or using excess enzymes may significantly reduce the final yield of the purified single-stranded DNA. Thus, it is highly recommended to follow the progress of the digestion reaction in a test run by taking out an aliquot of the reaction mixture every 10 min after initial 20 min incubation and analyzing them on the gel. Normally, the band of double-stranded DNA becomes lighter as the incubation time increases, indicating more double-stranded DNA has been converted into single-stranded DNA. Inactivation of lambda exonuclase by heating at 75°C for 10 min after digestion is recommended by most manufacturers. However, this step may be omitted if the digestion products are immediately subjected to agarose gel electrophoresis.

6. Lambda exonuclease is a highly processive 5' to 3' exodeoxyribonuclease that selectively digests the phosphorylated strand of double stranded DNA and has significantly reduced activity toward single-stranded DNA *(9)*. In many cases, lambda exonuclease cannot completely digest the phosphorylated strand of double-stranded DNA even using 10-fold excess of enzymes and longer incubation time. Thus, the single-stranded DNA must be purified from the gel after electrophoresis.
7. Since ethidium bromide binds less efficiently to ssDNA than to dsDNA, the ssDNA concentration estimated using gel electrophoresis and EZ load precision molecular mass ruler will always be lower than its real concentration.
8. A small aliquot (5 μL) of the reaction mixture should be taken and electrophoresed on an agarose gel before the purification step. The size of the cleaved DNA fragments is expected to be around 50 bp. If fragments larger than 100 bp are seen, add more DNase I and repeat the same digestion step.
9. Any commercially available thermostable DNA polymerase can be used to reassemble the DNA fragments. However, the use of DNA polymerases with proofreading activity such as *Pfu* DNA polymerase (Stratagene, La Jolla, CA), *Vent* DNA polymerase (New England Biolabs, Beverly, MA), or *Pfx* DNA polymerase (Invitrogen Life Technolgies, Carlsbad, CA) can minimize the introduction of point mutations in the chimeric progeny genes. When setting up reactions with these polymerases, it is very important to add the polymerase last since, in the absence of dNTPs, the 3'to 5'exonuclease activity of the polymerase can degrade DNAs.
10. The number of cycles depends on the fragment size. Assembly from small fragments may require more cycles than assembly from large fragments. Low annealing temperature may also be required for reassembling small fragments and for reassembling fragments derived from genes with low homology. The increasing elongation time for subsequent cycles correlates to the increasing fragment size as full- length chimeric genes are created.

11. *Pfu* DNA polymerase or other high fidelity DNA polymerases have a much slower processivity than *Taq* DNA polymerase, leading to lower extension rates. Processivity can be significantly improved (without significant loss in fidelity) by using a 1:1 ratio (1.25 Units each in a 100 μL reaction) of *Taq* DNA polymerase and *Pfu* DNA polymerase, rather than *Pfu* DNA polymerase alone ***(10)***. If this is done, an appropriate 10X reaction buffer for *Taq* DNA polymerase should be used in place of the 10X *Pfu* reaction buffer. It is noteworthy that certain commercial polymerase blends such as *Taq/Pwo* DNA polymerase (Expand High Fidelity PCR system from Roche Diagnostics) or *Taq*Plus Precision/*Pfu* (*Taq*Plus Precision PCR system of Stratagene) will yield similar results.

References

1. Crameri, A., Raillard, S. A., Bermudez, E., and Stemmer, W. P. (1998) DNA shuffling of a family of genes from diverse species accelerates directed evolution. *Nature* **391,** 288–291.
2. Ness, J. E., Welch, M., Giver, L., et al. (1999) DNA shuffling of subgenomic sequences of subtilisin. *Nat. Biotechnol.* **17,** 893–896.
3. Chang, C. C., Chen, T. T., Cox, B. W., et al. (1999) Evolution of a cytokine using DNA family shuffling. *Nat. Biotechnol.* **17,** 793–797.
4. Christians, F. C., Scapozza, L., Crameri, A., Folkers, G., and Stemmer, W. P. (1999) Directed evolution of thymidine kinase for AZT phosphorylation using DNA family shuffling. *Nat. Biotechnol.* **17,** 259–264.
5. Kumamaru, T., Suenaga, H., Mitsuoka, M., Watanabe, T., and Furukawa, K. (1998) Enhanced degradation of polychlorinated biphenyls by directed evolution of biphenyl dioxygenase. *Nat. Biotechnol.* **16,** 663–666.
6. Coco, W. M., Levinson, W. E., Crist, M. J., et al. (2001) DNA shuffling method for generating highly recombined genes and evolved enzymes. *Nat. Biotechnol.* **19,** 354–359.
7. Kikuchi, M., Ohnishi, K., and Harayama, S. (1999) Novel family shuffling methods for the in vitro evolution of enzymes. *Gene* **236,** 159–167.
8. Kikuchi, M., Ohnishi, K., and Harayama, S. (2000) An effective family shuffling method using single-stranded DNA. *Gene* **243,** 133–137.
9. Mitsis, P. G. and Kwagh, J. G. (1999) Characterization of the interaction of lambda exonuclease with the ends of DNA. *Nucleic Acids Res.* **27,** 3057–3063.
10. Zhao, H. and Arnold, F. H. (1997) Optimization of DNA shuffling for high fidelity recombination. *Nucleic Acids Res.* **25,** 1307–1308.

13

In Vitro DNA Recombination by Random Priming

Olga Esteban, Ryan D. Woodyer, and Huimin Zhao

1. Introduction

Variation coupled to selection is the hallmark of natural evolution. Although there is no full agreement concerning the best way to create variation, mutation or recombination *(1)*, computational simulation studies have demonstrated the importance of homologous recombination in the evolution of biological systems *(2,3)*. As compared to random mutagenesis, recombination may be advantageous in combining beneficial mutations that have arisen independently and may be synergistic, while simultaneously removing deleterious mutations.

The principles of natural evolution may be equally applied to the molecular evolution directed by an experimenter. For example, natural random mutagenesis is commonly simulated by the use of error prone PCR, while natural recombination is simulated by various in vitro recombination-based methods. These directed evolution methods have been extremely effective in engineering proteins, metabolic pathways, and whole genomes with improved or novel functions *(4–6)*. The power of in vitro recombination was first demonstrated by Stemmer in 1994 with the technique of DNA shuffling or sexual PCR *(7,8)*, in which DNA fragments generated by random digestion of parental genes with DNase I are combined and reassembled into full-length chimerical progeny genes in a PCR-like process. Since then, a number of in vitro recombination methods have been developed, including the staggered extension process (StEP) recombination *(9)*, random-priming in vitro recombination (RPR) *(10)* and random chimeragenesis on transient templates (RACHITT) *(11)*.

Here, we describe the method and protocol of random-priming in vitro recombination (RPR). As illustrated in **Fig. 1**, RPR relies on the reassembly of DNA fragments that are created by extension of random primers rather than by

From: *Methods in Molecular Biology, vol. 231: Directed Evolution Library Creation: Methods and Protocols*
Edited by: F. H. Arnold and G. Georgiou © Humana Press Inc., Totowa, NJ

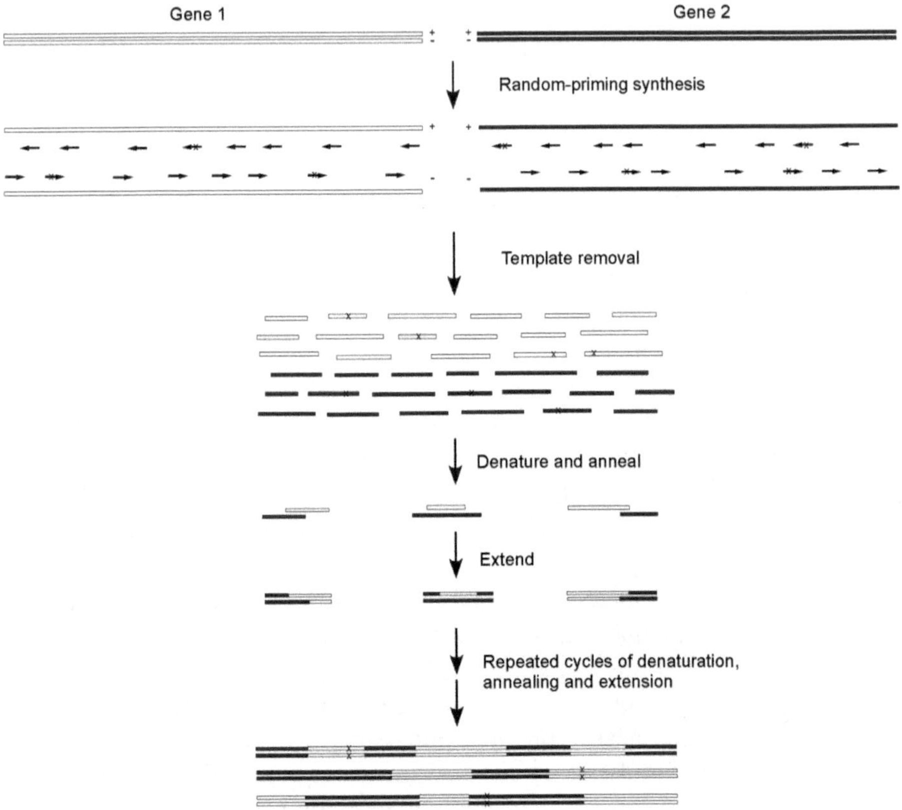

Fig. 1. Schematic of random priming in vitro recombination (RPR). For simplicity, only two DNA templates are shown. Random hexanucleotide primers are annealed to the templates and extended by Klenow fragment to yield a pool of different sized random extension products. After the removal of the oligonucleotides and the templates, the homologous fragments are reassembled into full-length chimerical genes in a PCR-like process. The full-length genes will be amplified by a standard PCR and subcloned into an appropriate vector.

DNase I fragmentation as in DNA shuffling. Random primers such as random hexamers are annealed to the template DNAs and then extended by a DNA polymerase at or below room temperature. The resulting DNA fragments are subsequently assembled into full-length genes by repeated thermocycling in the presence of a thermostable DNA polymerase. RPR has several potential advantages over DNA shuffling: (1) Since the random-priming DNA synthesis is independent of the length of the DNA template, DNA fragments as small as 200 nucleotides can be primed equally well as large molecules such as linearized plasmids. This offers unique potential for engineering peptides or ribozymes; (2) RPR allows the use of single-stranded DNA or RNA templates and requires significantly less parental DNA than DNA shuffling; (3) In RPR, mutations introduced by misincorporation and mispriming can further increase the sequence diversity created by recombination; (4) RPR overcomes the drawbacks of DNA shuffling associated with DNase I random fragmentation, such as the potential sequence bias during the gene reassembly process and the requisite complete removal of DNase I prior to reassembly.

2. Materials

1. DNA templates for random-priming synthesis (*see* **Note 1**).
2. $dp(N)_6$ random primers (Pharmacia Biotech Inc., Piscataway, NJ) (*see* **Note 2**).
3. Oligonucleotide primers for amplification of the recombined genes (*see* **Note 3**).
4. Klenow fragment of *E. coli* DNA polymerase I (Boehringer Mannheim, Indianapolis, IN) (*see* **Note 4**).
5. dNTP mix: 5 m*M* of each dNTP.
6. 10X reaction buffer: 900 m*M* HEPES, pH 6.6, 100 m*M* $MgCl_2$, 20 m*M* dithiothreitol (DTT).
7. 10X cloned *Pfu* PCR buffer: 200 m*M* Tris-HCl, pH 8.8, 100 m*M* KCl, 100 m*M* $(NH4)_2SO_4$, 20 m*M* $MgSO_4$, 1% Triton X-100, and 1 mg/mL bovine serum albumin (Stratagene, La Jolla, CA).
8. 10X *Taq* PCR buffer: 100 m*M* Tris-HCl, pH 9.0, 50 m*M* KCl and 0.1% Triton X-100 (Promega, Madison, WI).
9. Cloned *Pfu* DNA polymerase (Stratagene, La Jolla, CA).
10. *Taq* polymerase (Promega, Madison, WI).
11. Agarose gel electrophoresis supplies and equipment.
12. Microcon-100 filters and Microcon–3 or –10 filters (Amicon, Beverly, MA).
13. QIAquick PCR purification kit (Qiagen, Valencia, CA) or your favorite method.

3. Methods

1. Prepare DNA templates to be recombined using PCR amplification or restriction digestion of plasmids containing the target sequences.
2. Combine 0.7 pmol of the template DNAs with 7 nmol of $dp(N)_6$ random primers in a total volume of 65 μL (*see* **Note 5**).
3. Incubate the mixture at 100°C for 5 min and transfer onto ice immediately.
4. Add 10 μL of 10X reaction buffer, 10 μL of dNTP mix, 10 μL of distilled water, and 5 μL of the Klenow fragment of *E. coli* DNA polymerase I (2 U/μL) (*see* **Note 6**).
5. Incubate the reaction mixture at 22°C for 3 h and then terminate the reaction by placing the reaction tube on ice.
6. Run a small aliquot of the reaction mixture on an agarose gel. A faint, low molecular weight smear (50–500 bp) should be visible. If not, continue to incubate the reaction mixture at 22°C for an additional 2–3 h (*see* **Note 7**).

7. Add 100 μL of ice-cold deionized water to the reaction mixture and purify the random primed products from the templates, proteins, and large nascent DNA fragments by passing the whole reaction mixture through a Microcon-100 centrifugation filter at 500*g* for 10–15 min at room temperature (*see* **Note 8**).
8. Pass the flow-through fraction on either a Microcon–3 or –10 filter at 14,000*g* for 30 min at room temperature to remove the primers, fragments shorter than 30 nucleotides and unincorporated dNTPs. Recover the retentate fraction (~65 μL).
9. Combine 10 μL of purified fragments, 5 μL of 10X cloned *Pfu* buffer, 2 μL of dNTP mix (5 m*M* of each dNTP), and 0.5 μL of cloned *Pfu* polymerase (2.5 U/μL) and bring to a total volume of 50 μL in sterilized water (*see* **Note 9**).
10. Run the assembly reaction using the following program: 3 min at 96°C followed by 40 cycles of 30 s at 94°C, 1 min at 55°C, 1 min + 5 s/cycle at 72°C and finally 7 min at 72°C (*see* **Note 10**).
11. Run a small aliquot of the reaction mixture on an agarose gel. A smear extending through the size of the expected full-length gene should be visible. Otherwise, run an additional 10–20 cycles, or lower the annealing temperature.
12. Combine 1 μL of the assembly reaction, 0.3–1.0 μ*M* each primer, 10 μL of 10X *Taq* buffer, 10 μL of 2 m*M* each dNTP and 2.5 U *Taq/Pfu* (1:1) mixture in a total volume of 100 μL (*see* **Note 11**).
13. Run the PCR reaction using the following program: 96°C for 2 min, 25 cycles of denaturation at 94°C for 1 min, annealing at 55°C for 1 min, and elongation at 72°C for 1 min (*see* **Note 12**), followed by one final step of elongation at 72°C for 7 min.
14. Run a small aliquot of the reaction mixture on an agarose gel. In most cases, there is only a single band at the correct size. Purify the full-length PCR product using QIAquick PCR purification kit, or other reliable method. If there is more than one band, purify the product of correct size using QIAEX II gel purification kit.
15. Digest the product with the appropriate restriction endonucleases, and ligate to the desired cloning vector.

4. Notes

1. In most cases, double-stranded DNA templates are used, prepared from plasmids containing the target sequences using standard PCR techniques. However, random priming synthesis can also work for single-stranded DNA and RNA templates.
2. Random hexanucleotides are used for priming since they are long enough to form stable duplexes with the templates and short enough to ensure random annealing. While longer primers can also be used, annealing may not remain random for short genes.
3. Primer design should follow standard criteria including similar melting temperatures and elimination of self-complementarity or complementarity of primers to each other. Free computer programs such as Primer3 at Biology Workbench (http://workbench.sdsc.edu) can be used to design primers. Typically, primers should also include unique restriction sites for subsequent directional subcloning.
4. The Klenow fragment is the large fragment of *E. coli* DNA polymerase I, which lacks the 5' to 3' exonuclease activity of the full-length polymerase. There are

two main reasons for its use: 1) the lack of the 5' to 3' exonuclease activity will prevent the polymerase from removing bases from the primers, 2) it has a high activity at low temperatures where annealing can be less specific and shorter primers can be utilized (both enhancing random priming synthesis). Other polymerases lacking 5'-3' exonuclease activity may be substituted (*see* **Note 6**).

5. Since the average size of extension products is an inverse function of primer concentration ***(12)***, proper conditions for random-priming synthesis can be easily set for a given gene. For example, to obtain extension products with 50–500 bases, a 10,000-fold molar excess of primer over template should be used. The high primer concentration can facilitate production of shorter fragments, presumably by blocking extension.
6. Alternatively, other DNA polymerases such as bacteriophage T4 DNA polymerase, T7 Sequenase version 2.0 DNA polymerase, the Stoffel fragment of *Taq* polymerase and *Pfu* polymerase can be used during the random priming synthesis. When misincorporation is a concern, polymerases presenting 3' to 5'exonuclease (proofreading) activity such as bacteriophage T4 DNA polymerase and *Pfu* polymerase should be used. Bacteriophage T4 DNA polymerase, unlike Klenow fragment, has no strand displacement activity and therefore the isothermal amplification efficiency will be much reduced.
7. To better detect the low molecular weight smear, dye-free loading buffers with ~20% glycerol should be used for the samples and in the meantime, loading buffers with tracking dyes should be used in empty lanes to monitor the migration.
8. Do not exceed the recommended centrifugation speed since a significant amount of template may pass through the filter. A longer centrifugation time should be used if the suggested time is not adequate to purify the entire reaction mixture.
9. Any commercially available thermostable DNA polymerase can be used to reassemble the DNA fragments. However, the use of DNA polymerases with proofreading activity such as *Pfu* DNA polymerase (Stratagene, La Jolla, CA), *Vent* DNA polymerase (New England Biolabs, Beverly, MA), or *Pfx* DNA polymerase (Invitrogen Life Technolgies, Carlsbad, CA) can minimize the introduction of point mutations in the chimerical progeny genes. When setting up reactions with these polymerases, it is very important to add the polymerase last since, in the absence of dNTPs, the 3' to 5'exonuclease activity of the polymerase can degrade DNAs.
10. The number of cycles depends on the fragment size. Assembly from small fragments may require more cycles than assembly from large fragements. Small fragments may also require a lower annealing temperature, at least during the initial cycles. The increasing annealing time for subsequent cycles correlates to the increasing fragment size as full length chimerical genes are created.
11. *Pfu* DNA polymerase or other high-fidelity DNA polymerases have a much slower processivity than *Taq* DNA polymerase, leading to lower extension rates. Processivity can be improved significantly (without significant loss in fidelity) by using a 1:1 ratio (1.25 Units each in a 100 µL reaction) of *Taq* DNA polymerase and *Pfu* DNA polymerase rather than *Pfu* DNA polymerase alone ***(13)***. If this is done, an appropriate 10X reaction buffer for *Taq* DNA polymerase should

be used in place of the 10X *Pfu* reaction buffer. Certain commercial polymerase blends such as *Taq/Pwo* DNA polymerase (Expand High Fidelity PCR system from Roche Diagnostics) or *Taq*Plus Precision/*Pfu* (*Taq*Plus Precision PCR system of Stratagene) will yield similar results.

12. The elongation time depends on the gene size. For genes of 1 kb or less, an elongation time of 1 min will suffice. For genes larger than 1 kb, the elongation time should be adjusted accordingly.

References

1. Bäck, T. (1996) *Evolutionary Algorithms in Theory and Practice: Evolution Strategies, Evolutionary Programming, Genetic Algorithms.* Oxford University Press, New York, NY.
2. Holland, J. H. (1992) Genetic algorithms. *Sci. Am.* **267,** 66–72.
3. Forrest, S. (1993) Genetic algorithms: principles of natural selection applied to computation. *Science* **261,** 872–878.
4. Schmidt-Dannert, C. (2001) Directed evolution of single proteins, metabolic pathways, and viruses. *Biochemistry* **40,** 13,125–13,136.
5. Zhao, H. and Zha, W. (2002) Evolutionary methods for protein engineering, in *Enzyme Functionality: Design, Engineering and Screening*, (Svendsen, A., ed.) Marcel Dekker, New York, NY, in press.
6. Zhang, Y. X., Perry, K., Vinci, V. A., Powell, K., Stemmer, W. P., and del Cardayre, S. B. (2002) Genome shuffling leads to rapid phenotypic improvement in bacteria. *Nature* **415,** 644–646.
7. Stemmer, W.P. (1994) Rapid evolution of a protein in vitro by DNA shuffling. *Nature* **370,** 389–391.
8. Stemmer, W.P. (1994) DNA shuffling by random fragmentation and reassembly: in vitro recombination for molecular evolution. *Proc. Natl. Acad. Sci. USA* **91,** 10,747–10,751.
9. Zhao, H., Giver, L., Shao, Z., Affholter, J. A., and Arnold, F. H. (1998) Molecular evolution by staggered extension process (StEP) in vitro recombination. *Nat. Biotechnol.* **16,** 258–261.
10. Shao, Z., Zhao, H., Giver, L., and Arnold, F. H. (1998) Random-priming in vitro recombination: an effective tool for directed evolution. *Nucleic Acids Res.* **26,** 681–683.
11. Coco, W. M., Levinson, W. E., Crist, M. J., et al. (2001) DNA shuffling method for generating highly recombined genes and evolved enzymes. *Nat. Biotechnol.* **19,** 354–359.
12. Hodgson, C.P. and Fisk, R. Z. (1987) Hybridization probe size control: optimized 'oligolabelling'. *Nucleic Acids Res.* **15,** 6295.
13. Zhao, H. and Arnold, F. H. (1997) Optimization of DNA shuffling for high fidelity recombination. *Nucleic Acids Res.* **25,** 1307–1308.

14

Staggered Extension Process (StEP) In Vitro Recombination

Anna Marie Aguinaldo and Frances H. Arnold

1. Introduction

In vitro polymerase chain reaction (PCR)-based recombination methods are used to shuffle segments from various homologous DNA sequences to produce highly mosaic chimeric sequences. Genetic variations created in the laboratory or existing in nature can be recombined to generate libraries of molecules containing novel combinations of sequence information from any or all of the parent template sequences. Evolutionary protein design approaches, in which libraries created by in vitro recombination methods are coupled with screening (or selection) strategies, have successfully produced variant proteins with a wide array of modified properties including increased drug resistance *(1,2)*, stability *(3–6)*, binding affinity *(6)*, improved folding and solubility *(7)*, altered or expanded substrate specificity *(8,9)*, and new catalytic activity *(10)*.

Stemmer reported the first in vitro recombination, or "DNA shuffling," method for laboratory evolution *(11)*. An alternative method called the staggered extension process (StEP) *(12)* is simpler and less labor intensive than DNA shuffling and other PCR-based recombination techniques that require fragmentation, isolation, and amplification steps *(1,11,13,14)*. StEP recombination is based on cross hybridization of growing gene fragments during polymerase-catalyzed primer extension *(12)*. Following denaturation, primers anneal and extend in a step whose brief duration and suboptimal extension temperature limit primer extension. The partially extended primers randomly reanneal to different parent sequences throughout the multiple cycles, thus creating novel recombinants. The procedure is illustrated in **Fig. 1**. The full-length recombinant products can be amplified in a second PCR, depending on the product yield of the StEP reaction. The StEP method has been used to

From: *Methods in Molecular Biology, vol. 231: Directed Evolution Library Creation: Methods and Protocols*
Edited by: F. H. Arnold and G. Georgiou © Humana Press Inc., Totowa, NJ

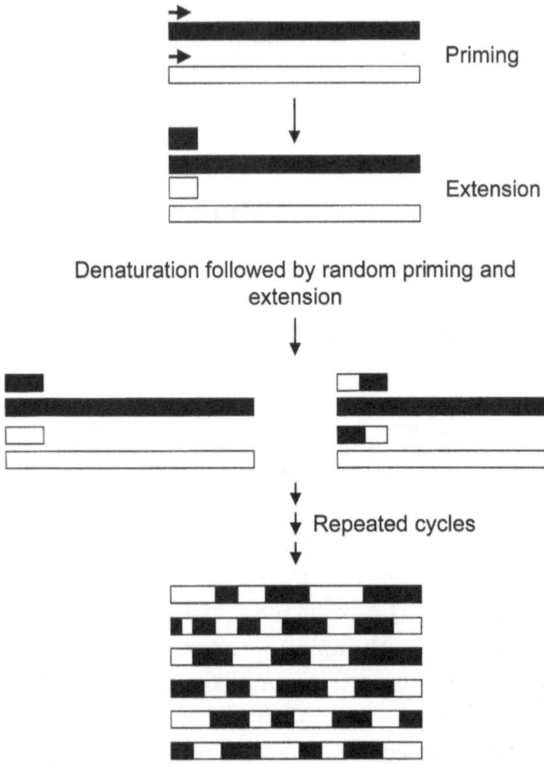

Fig. 1. StEP recombination, illustrated for two gene templates. Only one primer and single strands from the two genes (open and solid blocks) are shown for simplicity. During priming, oligonucleotide primers anneal to the denatured templates. Short fragments are produced by brief polymerase-catalyzed primer extension that is interrupted by denaturation. During subsequent random annealing-abbreviated extension cycles, fragments randomly prime the templates (template switching) and extend further, eventually producing full-length chimeric genes. The recombinant full-length gene products can be amplified in a standard PCR (optional).

recombine templates with sequence identity ranging from single base differences to natural homologous genes that are approx 80% identical.

2. Materials

1. DNA templates containing the target sequences to be recombined (*see* **Note 1**).
2. Oligonucleotide primers universal to all templates to be recombined (*see* **Note 2**).
3. *Taq* DNA polymerase (*see* **Note 3**).
4. 10X PCR buffer: 500 m*M* KCl, 100 m*M* Tris-HCl, pH 8.3.
5. 25 m*M* $MgCl_2$.
6. dNTP solution: 10 m*M* of each dNTP.

7. Agarose gel electrophoresis supplies and equipment.
8. *Dpn*I restriction endonuclease (20 U/μL) and 10X supplied buffer (New England Biolabs, Beverly, MA).
9. QIAquick gel extraction kit (Qiagen, Valencia, CA) or your favorite method.

3. Methods

1. Combine 1–20 ng total template DNA, 0.15 μ*M* each primer, 1X PCR buffer, 200 μ*M* dNTP mix, 1.5 m*M* $MgCl_2$, 2.5 U *Taq* polymerase, and sterile dH_20 to 50 μL. Set up a negative control reaction containing the same components but without primers (*see* **Note 4**).
2. Run the extension protocol for 80–100 cycles using the following parameters: 94°C for 30 s (denaturation) and 55°C for 5–15 s (annealing/extension) (*see* **Notes 5** and **6**).
3. Run a 5–10 μL aliquot of the reactions on an agarose gel to check the quality of the reactions (*see* **Note 7**). If a discrete band with sufficient yield for subsequent cloning is observed after the StEP reaction, and the size of the full-length product is clearly distinguishable and easily separated from the original starting templates, proceed to **step 8**.
4. If parental templates were purified from a *dam+ Escherichia coli* (*E. Coli*) strain (*see* **Note 1**), combine 2 μL of the StEP reaction, 1X *Dpn*I reaction buffer, 5–10 U *Dpn*I restriction endonuclease, and sterile dH_20 to 10 μL. Incubate at 37°C for 1 h (*see* **Note 8**).
5. Amplify the target recombinant sequences in a standard PCR using serial dilutions (1 μL of undiluted, 1:10, 1:20, and 1:50 dilutions) of the *Dpn*I reaction (or the StEP reaction if *Dpn*I digestion was not done). Mix 1X PCR buffer, 1.5 m*M* $MgCl_2$, 200 μ*M* dNTP mix, 20 μ*M* of each primer, 2.5 U *Taq* DNA polymerase, and sterile dH_2O to 100 μL.
6. Run the amplification reaction for 25 cycles using the following parameters: 94°C for 30 s, 55°C for 30 s, and 72 C for 60 s for each 1 kb in length.
7. Run a 10-μL aliquot of the amplified products on an agarose gel to determine the yield and quality of amplification (*see* **Note 9**). Select the reaction with high yield and low amount of nonspecific products.
8. Gel purify the desired full-length reaction product following the manufacturer's protocol in the QIAquick gel purification kit. Digest the purified fragment with the appropriate restriction endonucleases for ligation into the preferred cloning vector.

4. Notes

1. Appropriate templates include plasmids carrying target sequences, sequences excised by restriction endonucleases and PCR amplified sequences. Reactions are more reproducible as template size decreases because this reduces the likelihood of nonspecific priming. For example, three 8.5-kb plasmid templates containing different 1.7-kb target sequences were less efficient for StEP recombination than 3-kb restriction fragments containing the target sequences. Unusually large plasmids and templates should be avoided.

Short template lengths may also pose a problem when the size is indistinguishable in length from the desired product. Conventional physical separation techniques, such as agarose gel electrophoresis, cannot be used to isolate the reaction product from the template, resulting in a high background of nonrecombinant clones. To minimize parental templates that may contribute to background nonrecombinant clones, plasmids used for template preparations (both intact plasmids and restriction fragments containing target sequences excised from plasmids) should be isolated from a methylation positive *E. coli* strain, e.g., DH5α (BRL Life Technologies, Gaithersburg, MD) or XL1-Blue (Stratagene, La Jolla, CA). These *dam+* strains methylate DNA. *Dpn*I, a restriction endonuclease that cleaves methylated GATC sites, can then be used to digest parental templates without affecting the PCR products.

2. Primer design should follow standard criteria including elimination of self-complementarity or complementarity of primers to each other, similar melting temperatures (within approx 2–4°C is best), and 40–60% G + C content. Primers of 21–24 bases in length work well.
3. Other investigators have used Vent DNA polymerase instead of *Taq* DNA polymerase in StEP recombination ***(15)***. Vent DNA polymerase is one of several thermostable DNA polymerases with proofreading activity leading to higher fidelity ***(16)***. Use of these alternative polymerases is recommended for DNA amplification when it is necessary to minimize point mutations. In addition, the proofreading activity of high-fidelity polymerases slows them down, offering an additional way to increase recombination frequency ***(17)***. Vent polymerase, for example, is reported to have an extension rate of 1000 nucleotides/min and processivity of 7 nucleotides/initiation event as compared to the higher 4000 nucleotides/min and 40 nucleotides/initiation event of *Taq* DNA polymerase ***(18)***. Slower rates lead to shorter extension fragments and greater crossover frequency.
4. The negative control reaction should be processed in the same manner as the sample reactions for all the steps of the procedure. No product should be visible for the no primer control throughout the procedure. If bands are present in the negative control similar to the sample reactions, products of the sample reactions may be the result of template contamination resulting in nonrecombinant clones.
5. The annealing/extension times chosen are based on the number of crossover events desired. Shorter extension times as well as lower annealing temperatures lead to increased numbers of crossovers due to the shorter extension fragments produced for each cycle. The size of the full-length product determines the number of reaction cycles. Longer genes require a greater number of reaction cycles to produce the full-length genes. The annealing temperature should be a few degrees lower than the melting temperature of the primers.
6. The progression of the fragment extensions can be monitored by taking 10 μL aliquots of a duplicate StEP reaction at defined cycle numbers and separating the fragments on an agarose gel. For example, samples taken every 20 cycles from StEP recombination of two subtilisin genes showed reaction product smears with average sizes approaching 100 bp after 20 cycles, 400 bp after 40 cycles, 800 bp

after 60 cycles, and a clear discrete band around 1 kb (the desired length) within a smear after 80 cycles *(12)*. DNA polymerases currently used in DNA amplification are very fast. Even very brief cycles of denaturation and annealing provide time for these enzymes to extend primers for hundreds of nucleotides. For *Taq* DNA polymerase, extension rates at various temperatures are: 70°C, > 60 nt/s; 55°C, approx 24 nt/s; 37°C, approx 1.5 nt/s; 22°C, approx 0.25 nt/s *(19)*. Therefore, it is not unusual for the full-length product to appear after only 10–15 cycles. The faster the full-length product appears in the extension reaction, the fewer the template switches that have occurred and the lower the crossover frequency. To increase the ≠recombination frequency, everything possible should be done to minimize time spent in each cycle: selecting a faster thermocycler, using smaller test tubes with thinner walls, and, if necessary, reducing the reaction volume.

7. Possible reaction products are full-length amplified sequence, a smear, or a combination of both. Appearance of the extension products may depend on the specific–sequences recombined or the template used. Using whole plasmids may result in nonspecific annealing of primers and their extension products throughout the vector sequence, which can appear as a smear on the gel. A similar effect may be observed for large templates. If no reaction products are visible, the annealing/extension times and the temperature of the StEP reaction will need to be determined empirically. Try reducing the annealing temperatures as well as modifying the primer and/or template concentrations.
8. The background from non-recombinant clones can be reduced following the StEP reaction by *Dpn*I endonuclease digestion to remove methylated parental DNA (*see* **Note 1**). At this point you want to get rid of the DNA template that is still in your reaction mixture before proceeding to amplification to prevent carryover contamination
9. If the amplification reaction is not successful and you get a smear with a low yield of full-length sequence, reamplify these products using nested internal primers separated by 50–100 bp from the original primers.

References

1. Stemmer, W. P. (1994) DNA shuffling by random fragmentation and reassembly: in vitro recombination for molecular evolution. *Proc. Natl. Acad. Sci. USA* **81,** 10,747–10,751.
2. Crameri, A., Raillard, S. A., Bermudez, E., and Stemmer, W. P. C. (1998) DNA shuffling of a family of genes from diverse species accelerates directed evolution. *Nature* **391,** 288–291.
3. Giver, L., Gershenson, A., Freskgard, P. O., and Arnold, F. A. (1998) Directed evolution of a thermostable esterase. *Proc. Natl. Acad. Sci. USA* **95,** 12,809–12,813.
4. Zhao, H. M. and Arnold, F. H. (1999) Directed evolution converts subtilisin E into a functional equivalent of thermitase. *Protein Eng.* **12,** 47–53.
5. Miyazaki, K., Wintrode, P. L., Grayling, R. A., Rubingh, D. N., and Arnold, F. H. (2000) Directed evolution study of temperature adaptation in a psychrophilic enzyme. *J. Mol. Biol.* **297,** 1015–1026.

6. Jermutus, L., Honegger, A., Schwesinger, F., Hanes, J., and Pluckthun, A. (2001) Tailoring in vitro evolution for protein affinity or stability. *Proc. Natl. Acad. Sci. USA* **98,** 75–80.
7. Waldo, G. S., Standish, B. M., Berendzen, J., and Terwilliger, T. C. (1999) Rapid protein-folding assay using green fluorescent protein. *Nat. Biotech.* **17,** 691–695.
8. Zhang, J. H., Dawes, G., and Stemmer, W. P. C. (1997) Directed evolution of a fucosidase from a galactosidase by DNA shuffling and screening. *Proc. Natl. Acad. Sci. USA* **94,** 4504–4509.
9. Kumamaru, T., Suenaga, H., Mitsuoka, M., Watanabe, T., and Furukawa, K. (1998) Enhanced degradation of polychlorinated biphenyls by directed evolution. *Nat. Biotech.* **16,** 663–666.
10. Altamirano, M. M., Blackburn, J. M., Aguayo, C., and Fersht, A. R. (2000) Directed evolution of new catalytic activity using the alpha/beta-barrel scaffold. *Nature* **403,** 617–622.
11. Stemmer. W. P. C. (1994) Rapid evolution of a protein in vitro by DNA shuffling. *Nature* **370,** 389–391.
12. Zhao, H., Giver, L., Shao, Z., Affholter, J. A., and Arnold, F. H. (1998) Molecular evolution by staggered extension process (StEP) in vitro recombination. *Nat. Biotechnol.* **16,** 258–261.
13. Shao, Z., Zhao, H., Giver, L., and Arnold, F. H. (1997) Random-priming in vitro recombination: an effective tool for directed evolution. *Nucl. Acids Res.* **26,** 681–683.
14. Volkov, A. A. and Arnold, F. H. (2000) Methods for in vitro DNA recombination and random chimeragenesis. *Meth. Enzymol.* **28,** 447–456.
15. Ninkovic, M., Dietrich, R., Aral, G., and Schwienhorst, A. (2001) High-fidelity in vitro recombination using a proofreading polymerase. *BioTechniques* **30,** 530–536.
16. Cline, J., Braman, J. C., and Hogrefe, H. H. (1996) PCR fidelity of Pfu DNA polymerase and other thermostable DNA polymerases. *Nucleic Acids Res.* **24,** 3546–3551.
17. Judo, M. S. B., Wedel, A. B., and Wilson, C. (1998) Stimulation and suppression of PCR-mediated recombination. *Nucleic Acids Res.* **26,** 1819–1825.
18. Kong, H., Kucera, R. B., and Jack, W. E. (1993) Characterization of a DNA polymerase from the hyperthermophile archaea *Thermococcus litoralis.* Vent DNA polymerase, steady state kinetics, thermal stability, processivity, strand displacement, and exonuclease activities. *J. Biol. Chem.* **268,** 1965–1975.
19. Innis, M. A., Myambo, K. B., Gelfand, D. H., and Brow, M. D. (1988) DNA sequencing with Thermus aquaticus DNA polymerase and direct sequencing of polymerase chain reaction-amplified DNA. *Proc. Natl. Acad. Sci. USA* **85,** 9436–9440.

15

RACHITT

Gene Family Shuffling by Random Chimeragenesis on Transient Templates

Wayne M. Coco

1. Introduction

Random Chimeragenesis on Transient Templates (RACHITT) has been used to create libraries averaging 12 *(1)* or even 19 (L. Encell, unpublished) crossovers per gene in a single round of gene family shuffling. RACHITT creates chimeric genes by aligning parental gene "donor" fragments on a full-length DNA template *(1)*. The heteroduplexed top strand fragments are stabilized on the template by a single, long annealing step, taking advantage of full length binding by each fragment, rather than the binding of smaller overlaps, and by carrying out reactions at relatively high ionic strength. Fragments containing unannealed 5' or 3' termini are incorporated after flap trimming using the endo and exonucleolytic activities of *Taq* DNA polymerase and *Pfu* polymerase, respectively. After gap filling and ligation, the template, which was synthesized with uracils in place of thymidine, is rendered non-amplifiable by uracil-DNA glycosylase (UDG) treatment. Other methods of DNA shuffling by gene fragmentation and reassembly can result in reconstitution of one or all of the parental genes at unacceptably high frequencies in the final shuffled library *(2–4)*. Libraries with limited diversity owing to low numbers of crossovers can also result *(3,5,6)*. Both of these problems result because the hybridization of perfectly matching top and bottom strand fragments from the same gene are preferred over the heteroduplex hybridizations that lead to crossovers between genes *(3)*. The recent use of either top or bottom strands from each parental gene represents an improvement to sexual PCR, assuring at least one crossover per gene *(5)*. RACHITT similarly circumvents these issues, but by including a

From: *Methods in Molecular Biology, vol. 231: Directed Evolution Library Creation: Methods and Protocols*
Edited by: F. H. Arnold and G. Georgiou © Humana Press Inc., Totowa, NJ

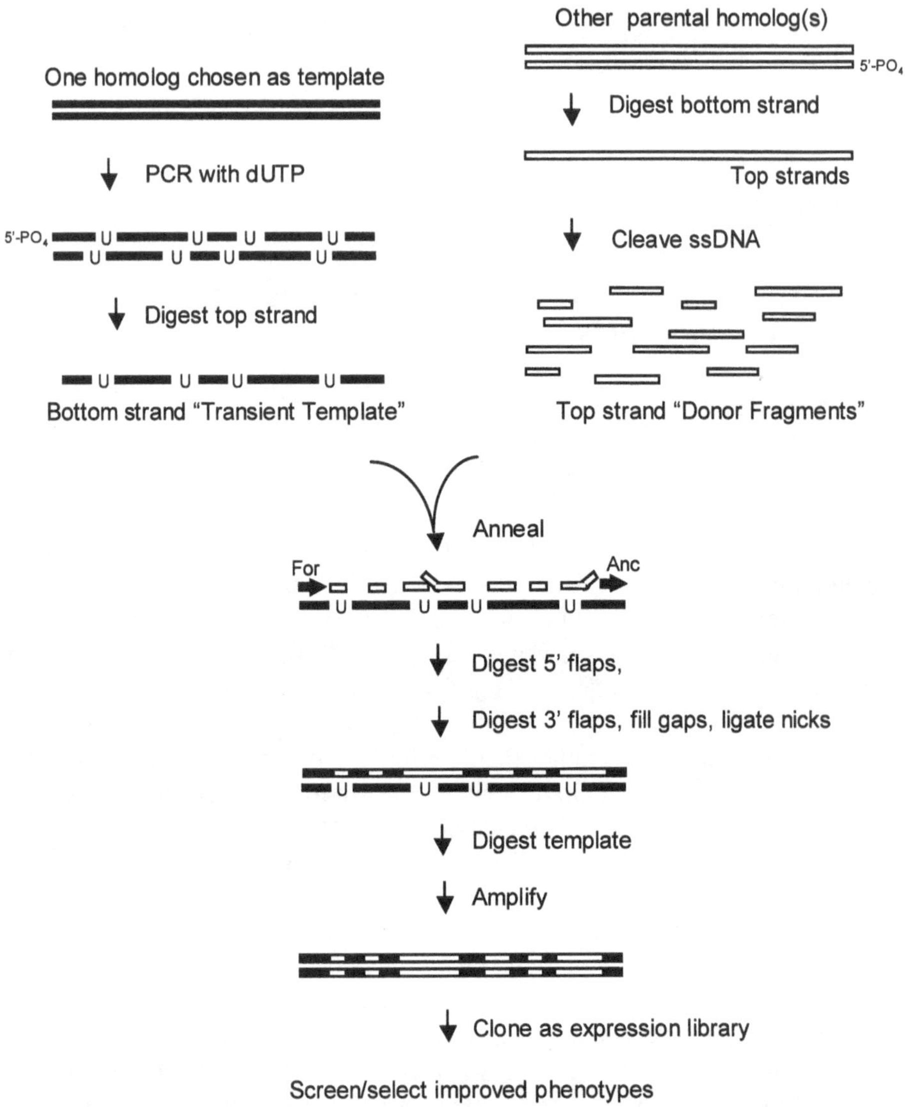

Fig. 1. RACHITT Overview. RACHITT begins with production of a single stranded bottom strand "Transient Template" containing uracil and production of single stranded top strand "Donor Fragments." The fragments are annealed to the template and joined to form a continuous chimeric top strand. Anchor oligonucleotide (Anc) protects the template 5'-terminus from the nucleases used to trim unannealed fragment flaps. Forward primer (For) protects the 3'-terminus and primes elongation to the first annealed fragment. The template is then degraded and the chimeric top strand amplified and cloned to result in a gene family shuffled library.

full-length bottom strand from the parental gene chosen as the template a nd only top strand "donor" fragments from the remaining parental genes (*see* **Fig. 1**). By this design, crossovers occur with virtually every top strand fragment incorporation event. Multiple crossovers thus occur at high frequency in RACHITT because multiple donor fragments from one or more parental genes can simultaneously anneal to any given template. RACHITT thus typically produces more diverse libraries. Improved diversity in recombined libraries increases exploration of the evolutionary fitness landscape, which is particularly important in the initial rounds of molecular evolution *(7,8)*. The novel process of layering multiple gene fragments onto a single template also allows unique flexibility in directing the desired representation of individual parental genes in the final chimeric library (*see* **Note 9**). It can also assist in library screening (*see* **Note 10**). Additional advantages and aspects of RACHITT were previously discussed *(1)*. The entire method can be completed in a few days.

2. Materials

1. Parental gene homologs to be shuffled.
2. *Taq* DNA polymerase.
3. 10X *Taq* DNA polymerase PCR buffer (with 15 m*M* $MgCl_2$).
4. Forward and reverse oligonucleotide primers with 5'-OH termini to amplify each parental gene.
5. Forward and reverse oligonucleotide primers with 5'-phosphate termini to amplify each parental gene (*see* **Note 12**).
6. Ultra Pure dNTPs (*see* **Note 2**) in sterile water or Tris-EDTA buffer (TE).
7. dUTP in sterile water or TE.
8. Agarose gel electrophoresis materials and apparatus, 1X Tris-Acetate EDTA (TAE) running buffer.
9. Centrifugal filtration-based spin columns (*see* **Note 3**).
10. Silica based DNA purification kit (*see* **Note 4**).
11. TE buffer: 10 m*M* Tris-HCl, pH 8.0, 1 m*M* EDTA. Sterilized.
12. Quartz cells (cuvets) for spectrophotometry at 260 nm.
13. Lambda exonuclease buffer, 10X: 670 m*M* glycine-KOH, pH 9.4, 25 m*M* $MgCl_2$, 500 μg/mL of bovine serum albumin (BSA).
14. Lambda exonuclease.
15. Centrifugal vacuum concentration apparatus (*see* **Note 5**).
16. *E. coli* tRNA, nuclease free. Dilute fresh from (0.1 mg/mL) stock for each use.
17. Uracil-DNA glycosylase (UDG).
18. 10X UDG buffer: 200 m*M* Tris-HCl, pH 8.0, 10 m*M* EDTA, 10 m*M* dithiothreitol (DTT).
19. Nuclease free bovine serum albumin (e.g., Ac-BSA as supplied for restriction endonuclease reactions by New England Biolabs, Inc.).
20. Gel filtration spin columns for recovery of 50 μL to 100 μL volumes (e.g., CentriSep, Princeton Separations).

21. Deoxyribonuclease I from bovine pancreas (DNase I). Prepare DNase I solution at 1 mg/mL in 0.01 *N* HCl. Divide into 5 μL aliquots and store at –70°C.
22. 10X DNase I buffer: Make fresh each day by combining in the following order (to reduce flocculation): 90 μL of water, 30 μL of 1 *M* $MnCl_2$, 30 μL of 10 mg/mL BSA, 150 μL of 1 *M* Tris-HCl, pH 7.4.
23. Polyacrylamide gel electrophoresis (PAGE) materials and apparatus, 1X TBE running buffer, pre-cast 6% PAGE-urea gels (*see* **Note 6**).
24. 2X TBE-urea PAGE sample buffer: 1X TBE, 4.2 g urea, 1.3 g ficoll, water to 10.0 mL.
25. Single stranded RNA size markers ranging from 50 nt or less to 200 nt or more.
26. "Anchor oligonucleotide" (*see* **Note 7**, **Fig. 1**).
27. *Taq* DNA ligase.
28. *Taq* DNA ligase buffer, 10X: 200 m*M* Tris-HCl, pH 7.6 at 25°C, 250 m*M* potassium acetate, 100 m*M* magnesium acetate, 100 m*M* DTT, 10 m*M* NAD, 0.1% Triton X-100 (after New England Biolabs, Inc., *see* **Note 8**). Keep frozen.
29. *Pfu* polymerase.

3. Methods

This section describes generation of transient template, generation of top strand "donor" fragments and chimeragenesis reactions, digestion of the template and cloning the chimeric library.

3.1. Generation of Transient Templates

We begin with the isolation of ssDNA bottom strand templates that contain uracil. Choice of which gene will be used as the template is the first step (*see* **Notes 9** and **10**). Next comes 1) incorporating uracil into the template gene by PCR, 2) digesting the unwanted top strand using lambda exonuclease and isolation of the transient template and 3) confirming the quality of the template.

3.1.1. Incorporating Uracil into the Template Gene

Generating a ssDNA bottom-strand template that includes uracil proceeds efficiently by PCR in the presence of dUTP, followed by enzymatic digestion of the top strand (*see* **Note 11**). The top strand of the amplified target must possess a 5'-phosphate in order to make it susceptible to degradation by lambda exonuclease. This is accomplished by using a forward primer containing a 5'-phosphate modification (*see* **Note 12**). The desired bottom strand template is protected from degradation because it lacks a 5'-phosphorylated terminus and is thus not a substrate for this nuclease (*see* **Note 1**). Uracil in the template will allow its ultimate inactivation by uracil-DNA glycosylase (UDG) after completion of the chimeragenesis reactions. Incorporate the needed uracils by two successive PCR amplifications in the presence of dUTP as follows (*see* **Table 1**).

Table 1
Components for PCR with dUTP

Volume (μL)	Component
—	Template DNA (1–10 ng)
5	10X PCR Buffer with 15 m*M* $MgCl_2$
1	2.5 m*M* each dNTP*
1	2.25 m*M* dUTP
3	5 μ*M* 5'-phosphate forward primer
3	5 μ*M* 5'-hydroxyl reverse primer
—	1.25 U *Taq* DNA polymerase
—	Water to 50 μL

See* **Note 13.

1. PCR with dUTP (*see* **Table 1**)
2. Separate 1/10th of the above PCR reaction by preparative agarose gel electrophoresis. Excise the gene-length product band and recover by centrifugal filtration.
3. Re-amplify 1/100th of the gel-purified product in the presence of dUTP as above.
4. Analyze 3 μL from the second amplification by agarose gel electrophoresis. If a single product is present, no gel purification is necessary (*see* **Note 15**).
5. Purify the remaining 47 μL of the non-gel purified second PCR product by a silica based DNA purification method. Elute in a final volume of 50 μL of TE. The products can be stored indefinitely at –20°C.
6. Quantitate by spectrophotometry at 260 nm. Expect 30–90 fmol/μL of dsDNA (~1–3 μg total).

3.1.2. Degrading the Non-Template Strand with Lambda Exonuclease

1. Digest 1 μg of the purified second PCR product from **Subheading 3.1.1.** in a reaction with 4 μL of 10X lambda exonuclease buffer, 10U lambda exonuclease, and water to 40 μL. Incubate at 37°C 1.5 h.
2. Excise the ssDNA product after preparative agarose gel electrophoresis (*see* **Fig. 2A**; *see* **Notes 14, 16,** and **17**). Recover DNA from the gel slice by centrifugal filtration.
3. Determine concentration by spectrophotometry at 260 nm (*see* **Note 18**). Bring to 40 fmol/μL by dilution with water or by centrifugal vacuum concentration. If significant concentration is required, a CentriSep desalting step may be necessary (buffer exchange vs 0.5X TE) before concentrating the sample. Store at –20°C or at –70°C for longer term storage (*see* **Note 19**). This is the purified transient template and is used directly in the RACHITT reaction (~40 fmol/reaction).

3.1.3. Confirming the Quality of the Transient Template

It is important to be sure you are going forward with a full-length ssDNA template that has sufficient uracils to allow complete inactivation. The first

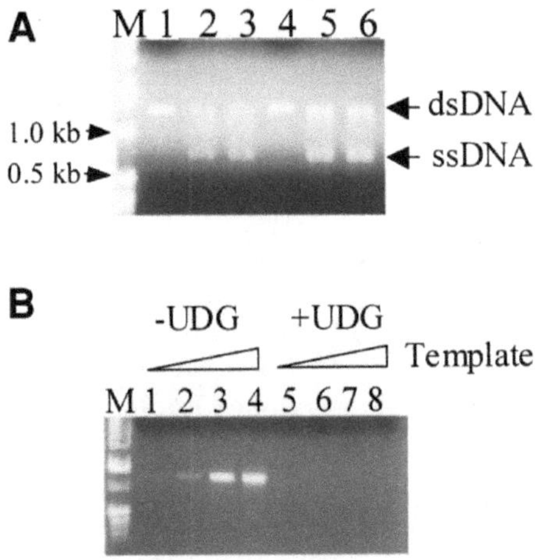

Fig. 2. Generation and confirmation of the Transient Template. **(A)** Lambda exonuclease reaction products. For a 1.2 kb gene, ssDNA migrates faster than dsDNA and is easily isolated (*see* **Note 16**). Lane M: DNA size marker; Lane 1 and 4: Two uncleaved dsDNA parental template genes amplified with dUTP. Lanes 2, 3 and 5, 6: the same genes after lambda exonuclease digestion. The reactions shown went to over 90% completion (*see* **Note 17**). **(B)** Confirming template length, polarity and transience. Lanes 1–4: 1:10,000, 1:1000, 1:100, and 1:10 dilutions of the isolated template after amplification by PCR. Lanes 5–8: As lanes 1–4, but digested with UDG prior to PCR amplification. The lack of product in 5–8 indicates the template will not be amplified after the RACHITT reaction and will thus not be cloned in the final library as contaminating parental genes.

of the following tests is strongly recommended. The second is useful if there is any question whether the excised band corresponds to the full-length ssDNA template.

1. Testing the template for transience and integrity: Serially dilute 0.5 μL of the transient template from **Subheading 3.1.2.** in 10-fold decrements from 1:10 to 1:10,000 with TE containing 10 μg/mL of tRNA. Prepare duplicate solutions containing 1 μL of each dilution, 1 μL of 10X UDG buffer, 1 μL of 1 mg/μL BSA, and 7 μL of water. To one sample of each duplicate, add 0.5 μL of UDG and incubate all samples for 30 min at 37°C. Amplify 2.5 μL of each sample by PCR (25 cycles) and analyze 10 μL by agarose gel electrophoresis. Quantity and integrity of the full-length template is confirmed by observing correct amplification products at or below the 1000-fold dilution where no UDG was used. Transience is confirmed by observing no PCR products even for the 1:10 dilution after UDG treatment (*see* **Fig. 2B**).
2. Confirming the transient template is ssDNA, is full length, and has correct polarity by primer extension: Use the minimum amount of the transient template needed

Table 2
Primer Extension on Transient Template

Volume (μL)	Component
—	Template DNA (10–20 ng)
5	10X PCR Buffer with 15 m*M* $MgCl_2$
1	2.5 m*M* each dNTPs
3	5 m*M* selected primer
—	1.25 U *Taq* DNA polymerase
—	Water to 50 μL

for detection by agarose gel electrophoresis. Prepare three reactions (*see* **Table 2**) using either the forward or reverse primer or no primer. Carry out a single primer extension reaction for each (e.g., 94°C for 2 min, annealing temp for 30 s, followed by 72°C for 1 min/kb). Use the same annealing temperature as in the PCR in **Subheading 3.1.1.** Analyze the resulting products by agarose gel electrophoresis. Only the forward primer should reconstitute the expected dsDNA band.

3.2. Generation of Donor Fragments

The parental genes chosen for use as top strand donor fragments are first made single stranded in a fashion analogous to the generation of template. Thereafter, however, they are fragmented using DNase I and size fractionated. Several micrograms of starting DNA are required.

3.2.1. Donor Gene Amplification

1. Perform sixteen 100 μL volume polymerase chain reactions (25 cycles) of each gene that will be used as a source of donor fragments (*see* **Table 3**). In contrast to the transient template PCR, here the reverse primer must contain a 5'-phosphorylated terminus and the forward primer must not contain a 5' phosphorylated terminus (*see* **Note 1**).
2. Pool identical reactions. Ethanol precipitate, wash, dry completely and resuspend in 100 μL of TE.
3. Determine concentration by spectrophotometry at 260 nm. Expected nominal yield is 25–125 μg. This should be considered a "nominal" yield, since some contaminating primers and nucleotides will remain after precipitation.

3.2.2. Removal of the Non-Donor Strand

Digest the unwanted bottom strand to leave the ssDNA top or "donor' strand by incubation at 37°C for 1.5 h with 10 μL of 10X lambda exonuclease buffer, 10 μg of the ethanol precipitated PCR product from **Subheading 3.2.1.**, 20 U lambda exonuclease and water to 100 μL. Inactivate enzyme at 75°C for 15 min. Carefully desalt by gel filtration spin cartridge equilibrated in TE. Analyze

Table 3
Donor Gene PCR

Volume (μL)	Component
—	Donor gene DNA (1–50 ng)
10	10X PCR Buffer with 15 m*M* $MgCl_2$
2	2.5 m*M* ultra pure dNTPs*
6	5 m*M* 5'-hydroxyl forward primer
6	5 m*M* 5'-phosphate reverse primer
—	2.5 U *Taq* DNA polymerase
—	Water to 100 μL

See* **Note 2.

1/20 volume by agarose gel electrophoresis to confirm the exonuclease went to about 80% completion or better (*see* **Note 20**).

3.2.3. DNase I Cleavage of the Donor Strand

The following was modified from a DNA sequencing protocol in **ref. *11***.

1. Dilute enzyme just before use into pre-chilled 1X DNase I buffer on ice. The following dilution is 1:200 (*see* **Note 21**): 20 μL of 10X DNase I Buffer, 179 μL of water, 1 μL of DNase I (from a frozen aliquot, thawed on ice, *see* **Materials**).
2. DNase I cleavage reaction: In a PCR tube, add the desalted, lambda exonuclease-treated PCR product from **Subheading 3.2.2.**, 10 μL of 10X DNase I Buffer and water to 98.5 μL. Bring to 15°C in a thermocycler or heat block; Add 1.5 μL of Diluted DNase I and mix well; After 8 min at 15°C, stop reaction with 5 μL of 0.5 *M* EDTA and place on ice; Bring thermocycler or heat block to 75°C. Remove samples from ice and heat 75°C, 15 min.
3. Gel purify by preparative TBE-urea denaturing PAGE as follows (*see* **Note 22**): Carefully desalt by gel filtration spin cartridge equilibrated with 0.5X TE. Concentrate by vacuum centrifugation to less than 10 μL. Bring volume to 10 μL with water. Add an equal volume of TBE-urea PAGE sample buffer. Add an equal volume of sample buffer to 2 μg of RNA size markers. Heat samples and markers to 90°C for 3 min. Load and run at 180 V for 30–50 min until the bromophenol blue dye has nearly traversed the gel. Stain with ethidium bromide in a clean tray. Visualize for the minimum time possible at 302 nm or a longer wavelength (*see* **Fig. 3**). Cut out the portion of the DNA smear corresponding to fragments migrating from 45–200 bases (and/or other ranges, if desired). Place gel slice in a 1.5-mL centrifuge tube and crush against the tube wall with a 1-mL disposable pipettor tip. Rinse tip clean of adhering polyacrylamide while adding 500 μL of TE to the crushed gel. Heat 75°C for 15 min to inactivate any contaminating nucleases. Allow DNA to elute from the gel at 4°C, over night (or at 37°C

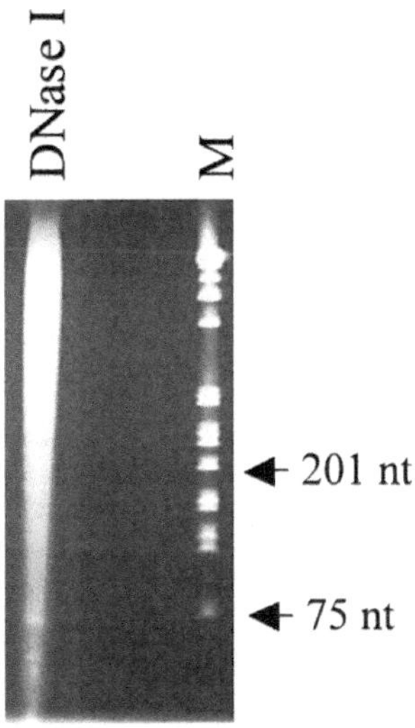

Fig. 3. DNase I cleavage reaction products. A. ssDNA donor gene was fragmented by DNase I digestion and separated by denaturing PAGE (DNase I). The desired fragment size range is excised by referencing the nucleotide size marker (M).

for 3 h with agitation). Spin down gel for 5 min at 10,000*g* or greater. Transfer the resulting supernatant to a clean 1.5-mL centrifuge tube. Evaporate to less than 50 µL by vacuum centrifugation. Bring volume to 50 µL with water. Desalt and exchange buffer by CentriSep gel filtration spin column equilibrated with 0.5X TE. Concentrate to less than 10 µL by centrifugal vacuum concentration and bring to 10 µL with water. Determine concentration of 1 µL of sample in water by spectrophotometry (*see* **Note 23**). Expect about 200 fmol/µL (*see* **Note 24**). Store at –20°C, or at –70°C for longer-term storage (*see* **Note 19**). This is the final Donor Fragment solution and is used directly in the RACHITT reaction.

3.3. RACHITT Reactions

With the isolation of donor fragments and transient template completed, the next step is the creation of chimeric genes. This proceeds by annealing fragments to the template, trimming non-annealed 5' flaps and ligating any fragments that happen to abut. Thereafter, 3' flaps are trimmed, gaps are filled and the resulting nicks are ligated. The transient template is then rendered non-amplifiable by treatment with UDG. PCR amplification of the chimeric top strands then results in a quantity of dsDNA genes suitable for efficient cloning to generate the final combinatorial library. Inclusion of two control reactions is also useful. One includes template, but no donor fragments; the other has fragments, but no template.

Table 4
Flap Endonuclease and Ligation Reaction

Components	Concentration	1	2	3
H_2O to 7.5 µL		3.5	3.5	2.5
E. coli tRNA	0.01 mg/mL, freshly diluted	1	1	1
Taq DNA ligase buffer	10 X	1	1	1
Transient template	40 fmol/µL	1	0	1
Forward primer*	5 pmol/µL	0.5	0.5	0.5
5'-phosphorylated anchor oligonucleotide*	5 pmol/µL	0.5	0.5	0.5
Donor Fragments	200 fmol/µL total	0	1	1

See* **Note 7.

3.3.1. Donor Fragment Annealing, Flap Endonuclease Trimming, and Ligation

1. If donor fragments from more than one parent are to be used, mix them in a 1:1 ratio for a final total concentration of 200 fmol/µL.
2. Add the appropriate components to three PCR tubes (*see* **Table 4**). After component addition, denature each sample at 94°C for 3 min, and then anneal at 65°C for 30 min.
3. Prepare fresh Cold Start Mix as a master mix containing 0.5 U of *Taq* DNA polymerase, 2 U of *Taq* DNA ligase, 0.44 µL of 10X *Taq* DNA ligase buffer, and water to 2.5 µL (multiply volumes for the total number of samples to be performed). Add 2.5 µL of Cold Start Mix to each tube at 65°C, and mix immediately by pipetting. Samples are stopped after 2 min by cooling on ice, then by freezing at –70°C.

3.3.2. Completing the Chimeric Strand: Gap Filling and Nick Ligation

Thaw each sample and dilute with 12 µL of water. Add 1 µL of each sample to three tubes containing the listed components (*see* **Table 5**) and save the remaining unused portion at –20°C. The reaction is started at 45°C and ramped to 65°C over 30 min.

3.3.3. Inactivation of the Transient Template

After completion of the chimeric strand, the transient template is no longer needed. Mix Template Inactivation Mixture components (*see* **Table 6**) and distribute 8 µL into one PCR tube for each reaction. A 2-µL sample of each Gap Filling and Ligation reaction from **Subheading 3.3.2.** is then carefully added to the Inactivation Mixture in each tube, taking care not to touch the sides of the tube with the pipettor tip (*see* **Note 25**). Incubate at 37°C for 30 min.

Table 5
Gap Filling and Ligation

Volume (μL)	Component
1	Diluted sample
1	10X *Taq* DNA Ligase buffer
0.1	10 m*M* ultra pure dNTPs*
—	1.25 U *Pfu* Polymerase
—	0.5 U *Taq* DNA polymerase
—	2.5 U *Taq* DNA ligase
—	Water to 10 μL

See* **Note 2.

Table 6
Template Inactivation Mixture

Volume (μL)	Component
1	10X UDG buffer
0.1	10 mg/mL BSA
—	0.5 U UDG
6.4	Water to 8 μL

3.3.4. Amplification, Cloning, and Analysis of the Chimeric Products

1. Amplify 2.5 μL of the chimeric sample from **Subheading 3.3.3.** in a 100-μL reaction, using up to 30 cycles of PCR.
2. Analyze 10 μL by agarose gel electrophoresis. Expect a single product with little smearing (*see* **Fig. 4** and **Note 26**).
3. Purify and clone the remaining 90 μL.
4. Test for successful chimeragenesis by sequencing four to 20 or more clones (*see* **Notes 27** and **28**).
5. If this initial library is satisfactory, it can be used directly for expression and screening of improved or novel phenotypes. Alternatively, the remaining unused samples from **Subheadings 3.3.1.–3.3.3.** can be processed in the remaining steps in order to generate larger libraries (*see* **Note 29**).

4. Notes

1. A 5'-OH group generally terminates the non-phosphorylated primer, but other non-phosphate modifications may also be acceptable.
2. Ultra pure nucleotide triphosphate solution is used to minimize unintentional incorporation of contaminating uracils in the final donor fragments. Uracils in the

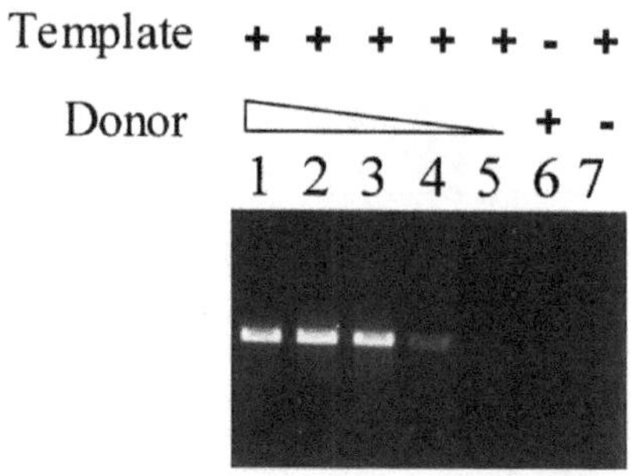

Fig. 4. Final RACHITT products and controls after PCR. Lanes 1–5 contain PCR products after RACHITT reactions with 214 fmol, 107, 42.8, 8.56, and 2.14 fmol of Donor fragments, respectively. Lane 6 and 7 are negative controls containing donor fragments, but no template or template, but no donor fragments, respectively.

final chimeric strand will be hydrolyzed by UDG and reduce the yield of chimeric genes. We have successfully used "PCR grade nucleotide mix" from Roche.

3. In steps where no buffer exchange, concentration or additional purification is required, DNA can be quickly recovered from gel slices by centrifugal filtration. Spin columns are available that retain gelled agarose, but allow ssDNA or dsDNA of any size to pass through (e.g., GenElute Agarose Spin Column, Sigma-Aldrich). Where desired, a silica based purification method can be used in place of centrifugal filtration throughout. Where silica based methods are substituted for the isolation of ssDNA, however, problems with quantitation may result (*see* **Note 4**).
4. Any silica based DNA purification kit or method should be satisfactory (e.g., Ultra Gel Clean, MoBio, Inc.). DNA purified by some such methods, however, can contain contaminants that absorb at 260 nm. Where DNA quantitation by spectrophotometry is in question, agarose gel electrophoresis can be used as an alternative method for quantitation of dsDNA. Keep in mind, however, that ssDNA can not be quantified directly using dsDNA markers on ethidium bromide stained gels (*see* **Note 17**).
5. At each centrifugal vacuum concentration step, it will not harm sample quality if the evaporation proceeds to completion. If a centrifugal vacuum concentrator (e.g., SpeedVac, Thermo Savant) apparatus is unavailable, other concentration methods should be equally acceptable.
6. Novex pre-cast TBE-urea 6% PAGE gels (Invitrogen, Inc.) were used for convenience. Self-made gels should work equally well, if tetramethylethylenediamine (TEMED) and ammonium persulfate (APS) are not used in excess, and are given time to completely react.
7. Adding a forward primer in the annealing step protects the template 3'-terminus from exonucleolytic cleavage and allows priming of elongation to the first annealed fragment in subsequent reactions. The "anchor" oligonucleotide pro-

tects the template 5'-terminus (*see* **Fig. 1**). Since the anchor oligonucleotide will be ligated after extension of an upstream donor fragment, the anchor oligonucleotide must possess a 5'-phosphate modification. Both forward primer and anchor oligonucleotides should have annealing temperatures for PCR of at least 60°C, in order to assure the maximal number of templates are bound by both. The higher ionic strength of the RACHITT reactions assures these primers will bind during the 65°C steps of the protocol. The forward primer can be the PCR primer used to amplify the parental genes. We have used forward primer: 5'-CCGTCGCCCTCGATCAGTG (T_m=68°C) and anchor oligonucleotide: 5'-PO4-GTTTCCAGCCACACCACCC (T_m=61°C). If desired, the complements of these sequences can be added to the termini of the template by standard techniques (e.g., by amplifying the template gene using tailed primers). All parental genes should be examined to assure there are no internal binding sites for these oligonucleotides on the strands to be used in the RACHITT reaction. Where other forward and anchor oligonucleotides will be designed, it is advisable to also check them for lack of secondary structure. If cloning will be carried out using restriction sites (recommended), they can be included in the primer tails or on each parental gene internal to the primer binding sites.

8. The buffer included with *Taq* DNA ligase from New England Biolabs, Inc. (NEB) is used for several reactions containing other enzymes. Buffers supplied by other providers have not been tested. If *Taq* DNA ligase is acquired from another source, it may be advisable to use a buffer made to the specifications of the buffer supplied by NEB (*see* **Materials**).
9. Alternative methods can have significant bias in the relative incorporation of fragments from various parental genes *(3)*. The ability to choose one gene as the sole template gene, and thus force its incorporation, represents a mechanism to reduce bias that is unique to RACHITT. For example, suppose three genes, A, B, and C, are to be recombined. Suppose gene A is 90% identical to gene B, and that gene C is 80% identical to both A and B. In such a case, C is the clear choice for the template, since A and B should allow similar numbers of incorporated fragments during RACHITT. If A were chosen as template, one would expect more contributions from the more similar gene, B, than from C. (Indeed, in some cases such bias may be desirable.) Further, in contrast to sexual PCR, increasing or decreasing the relative concentrations of donor fragments from different parents will alter their relative representation in the final library without increasing the number of unrecombined parental genes. Finally, altering the average fragment size and absolute concentration of donor fragments allows one to vary the representation of template sequences (incorporated by gap filling) relative to donor sequences (i.e., by varying the size and number of fragments on the average template and the gap size between fragments). Genes of 1.2 kb have worked well with this procedure. Much larger genes may have to be divided and shuffled in parts.
10. Choice of template can also improve ease of selection. Where only two parents are used, or where all parent genes have similar percent DNA sequence identities, the choice of template may take advantage of differing parental phenotypes

in the chosen library screening strategy. Although we did not find any contaminating parental genes in the dozens of recombinants we examined, they may exist in numbering numbers in experiments where the various parents are more divergent. In such cases, one should select as template, where possible, a parental gene with a phenotype or genotype that is not detected as a positive in the desired selection or screen. This strategy is made possible by the fact that, in RACHITT, donor gene fragments cannot regenerate an unrecombined parental gene.

11. It is a misconception *(9)* that RACHITT requires the use of phagemid vectors for the generation of ssDNA template.
12. An alternative to including 5'-phosphate modification during oligonucleotide synthesis is to perform a kinase reaction on the primer of interest using T4 polynucleotide kinase. We have occasionally found that a nominally redundant kinase step on oligonucleotides synthesized with 5'-phosphates increases the efficiency of lambda exonuclease strand degradation. A third option includes amplification with unmodified primers, followed by restriction endonuclease cleavage near one end of the amplicon to yield the appropriate 5'-phosphate. Keep in mind that the primer used for the initial amplification will have a truncated or absent binding site after restriction cleavage.
13. For small genes, it is advisable to use a nucleotide mix with 2.5 m*M* dGTP, dATP, and dCTP, but less or no dTTP. This may be necessary to provide enough uracil incorporation to ensure complete inactivation of the template in the uracil-DNA glycosylase step.
14. Single stranded DNA is sensitive to nucleases to which dsDNA is resistant. In all steps calling for preparative electrophoresis, gel apparatus should be cleaned and filled with unused running buffer. Preparative gels should be handled using only clean materials or rinsed gloves and should be stained with unused ethidium bromide solutions and rinsed or destained in clean water. DNA should be excised from gels without excess agarose.
15. Where multiple products are visible, but the desired product can be excised without contamination, the PCR reaction should be gel purified before proceeding. Where excision of a single product is not possible, the PCR reaction should be optimized. In our experience, where PCR yields a specific band without dUTP inclusion, no reoptimization has been necessary.
16. We typically use 0.8% agarose gels in 1X TAE. For a 1.2 kb gene, the single strand migrates at an apparent MW of about half the dsDNA gene. For very small genes, the ssDNA can migrate slower than dsDNA. Single- and double-stranded DNA can co-migrate for genes of around 800 bp. In such cases, after the lambda exonuclease reaction, you may have to cut the dsDNA with a restriction enzyme, use other gel buffers, or rely on other methods to separate the two species.
17. After lambda exonuclease digestion, the dsDNA and ssDNA bands may fluoresce with equal intensity. This is not a cause for concern, since ssDNA-ethidium bromide complexes fluoresce 10- to 20-fold less intensely than stained dsDNA. The yields of the desired ssDNA in such cases are entirely appropriate for the downstream reactions.

18. Use the extinction coefficient for ssDNA (8.5 liter·$cm^{-1}mmol^{-1}L^{-1}$, where L = length of the fragment in nucleotides).
19. We have anecdotally observed poorer performance from some single stranded templates after long-term storage. The same may apply to the ssDNA donor fragments.
20. The carryover of some residual dsDNA into the DNase I reactions will not affect the quality of the final library. If reducing the levels of dsDNA is desired, the ssDNA can be first gel purified as described for the ssDNA template. Alternatively, a ssDNA specific endonuclease, such as mung bean nuclease, can be used in place of DNase I in **Subheading 3.2.3.**
21. DNase I overdigestion can decrease the randomness of fragment lengths and endpoints. Underdigestion by DNase I will result in lower product yield in the size range desired. The activity of DNase I preparations can vary. If the 1:200 dilution results in over- or underdigestion, determine an appropriate dilution as follows. Scale down the Donor strand cleavage reaction and test a range of dilutions of DNase I. Use the dilution that gives significant product in the size range of interest after 8 min using the conditions from **Subheading 3.2.3.**, **step 2**.
22. DNase I products are separated on a denaturing gel to accurately recover the size range desired. There can be significant differences in the migration of similarly sized ssDNAs on non-denaturing gels.
23. Upon dilution of sample into a cuvette, this measurement will be near the limits of detection without an added fluorescent dye. If no dye is used, make sure to establish a stable baseline reading with water at 260 nm, add the sample to this cuvet, read the sample several times, and use the average value minus baseline. Alternatively, OliGreen (Molecular Probes, Inc.) can be added and fluorescence measured according to the manufacturer's specifications.
24. For convenience, concentration is calculated using the average length from the size range excised.
25. Sample left on the tube sides will contain undegraded template that can be transferred and amplified in the subsequent PCR.
26. Observation of products in the absence of donor fragments indicates read-through from the forward primer, but does not indicate a problem with the donor-containing reactions. When donor fragments occupy most templates, read-through products are suppressed. The "no-donor" reaction controls solely for proper polymerization on a viable template and for the subsequent PCR amplification.
27. The number of spontaneous point mutations should be consistent with the number of PCR cycles used to prepare template and donor fragments and to amplify the final library (that is, fewer than 1 mutation per 1000 bp in the final library) ***(1)***. This number can be further reduced by performing these preparative PCR reactions with a proofreading polymerase (e.g., *Pfu* or *Pwo* DNA polymerases) in place of *Taq* DNA polymerase. However, note that the dUTP incorporation step can be less efficient, depending on polymerase choice ***(10)***. Substitution of polymerases in the chimeragenesis reactions (**Subheadings 3.3.1.–3.3.2.**) has not been tested and should be attempted with caution.

28. In the final library, there should be no observed identical sibling clones and no non-chimeric parental genes in a random sampling of 20 clones. There should also be no insertions or deletions. For a single round of RACHITT, calculate the crossover density or average number of crossovers per gene *(1)*, the number of crossovers within regions of little homology *(1)*, and the relative absence of hotspots and cold spots for crossovers between each polymorphism. In comparing alternative conditions or alternative DNA shuffling methods, it is important to compare similar experiments. That is, shuffling the same number of genes that differ by the same percent sequence identity. Assemble data for the same number of randomLy chosen clones. Of course, comparing reactions using identical parental genes is ideal.
29. The protocol as described will provide an "initial" library of up to 55 million unique chimeric clones (maximum theoretical yield), enough for most high throughput screens. If the unused portion of each reaction is also processed, that maximum will increase to over 10^{10} unique genes. When processing the entire reaction, a concentration and buffer exchange (e.g., by silica-based DNA purification or by vacuum centrifugation and gel filtration spin columns) after **Subheading 3.3.3.** will help to reduce the final number of necessary PCRs. After concentration, fewer PCR cycles will be necessary.

Acknowledgments

I thank Daniel J. Monticello, Charles H. Squires and the other scientists of Enchira Biotechnology Corporation for their support and suggestions during development of the RACHITT method.

References

1. Coco, W. M., Levinson, W. E., Crist, M. J., et al. (2001) DNA shuffling method for generating highly recombined genes and evolved enzymes. *Nat. Biotechnol.* **19,** 354–359.
2. Kikuchi, M., Ohnishi, K., and Harayama, S. (1999) Novel family shuffling methods for the in vitro evolution of enzymes. *Gene* **236,** 159–167.
3. Moore, G. L., Maranas, C. D., Lutz, S., and Benkovic, S. J. (2001) Predicting crossover generation in DNA shuffling. *Proc. Natl. Acad. Sci. USA* **98,** 3226–3231.
4. Joern, J. M., Meinhold, P., and Arnold, F. H. (2002) Analysis of shuffled gene libraries. *J. Mol. Biol.* **316,** 643–656.
5. Kikuchi, M., Ohnishi, K., and Harayama, S. (2000) An effective family shuffling method using single-stranded DNA. *Gene* **243,** 133–137.
6. Hansson, L. O., Bolton-Grob, R., Massoud, T., and Mannervik, B. (1999) Evolution of differential substrate specificities in Mu class glutathione transferases probed by DNA shuffling. *J. Mol. Biol.* **287,** 265–276.
7. Voigt, C. A., Mayo, S. L., Arnold, F. H., and Wang, Z. G. (2001) Computational method to reduce the search space for directed protein evolution. *Proc. Natl. Acad. Sci. USA* **98,** 3778–3783.

8. Voigt, C. A., Kauffman, S., and Wang, Z. G. (2001) Rational evolutionary design: the theory of in vitro protein evolution. *Adv. Protein Chem.* **55,** 79–160.
9. Kurtzman, A. L., Govindarajan, S., Vahle, K., Jones, J. T., Heinrichs, V., and Patten, P. A. (2001) Advances in directed protein evolution by recursive genetic recombination: applications to therapeutic proteins. *Curr. Opin. Biotechnol.* **12,** 361–370.
10. Slupphaug, G., Alseth, I., Eftedal, I., Volden, G., and Krokan, H. E. (1993) Low incorporation of dUMP by some thermostable DNA polymerases may limit their use in PCR amplifications. *Anal. Biochem.* **211,** 164–169.
11. Sambrook, J., Fritsch, E. F., and Maniatis, T. (1989) *Molecular Cloning, Second Edition.* Cold Spring Harbor Laboratory Press, Plainview, NY.

16

The Creation of ITCHY Hybrid Protein Libraries

Marc Ostermeier and Stefan Lutz

1. Introduction

Incremental truncation is a method for creating a library of every one base truncation of dsDNA. Incremental truncation libraries can be created by time dependent Exo III digestions *(1)* or by the incorporation of α-phosphorothioate dNTPs (αS-dNTPs) *(2)*. The fusion of two incremental truncation libraries is called an ITCHY library *(3)*. An ITCHY library created from a single gene consists of genes with internal deletions and duplications. An ITCHY library created between two different genes consists of gene fusions created in a DNA-homology independent fashion. ITCHY libraries, as well as an incremental truncation-like method called SHIPREC *(4)*, have the potential to create proteins with improved or novel properties as well as to generate artificial families for in vitro recombination in a method called SCRATCHY *(5)* (*see* **Chapter 17**).

To create an ITCHY library, the two genes or gene fragments are cloned in tandem. In time-dependent truncation (*see* **Fig. 1**), the DNA is subjected to Exo III digestion starting from a unique restriction enzyme site between the two genes (RE1). During Exo III digestion, whose rate is control by the addition of NaCl, small aliquots are removed frequently and quenched by addition to a low pH, high salt buffer. Blunt ends are prepared by treatment with a single-strand nuclease and a DNA polymerase. This synchronized truncation library is uncoupled by digestion at a second restriction enzyme site (RE2) followed by religation. In truncation using αS-dNTPs (*see* **Fig. 2**), the entire plasmid is amplified by PCR using dNTPs and a small amount of αS-dNTPs. Subsequent digestion with Exo III is prevented from continuing past the randomly incorporated αS-dNMP *(6)*. Blunt ends are prepared by treatment with a single-strand nuclease and a DNA polymerase followed by unimolecular ligation to recyclize the vector.

From: *Methods in Molecular Biology, vol. 231: Directed Evolution Library Creation: Methods and Protocols*
Edited by: F. H. Arnold and G. Georgiou © Humana Press Inc., Totowa, NJ

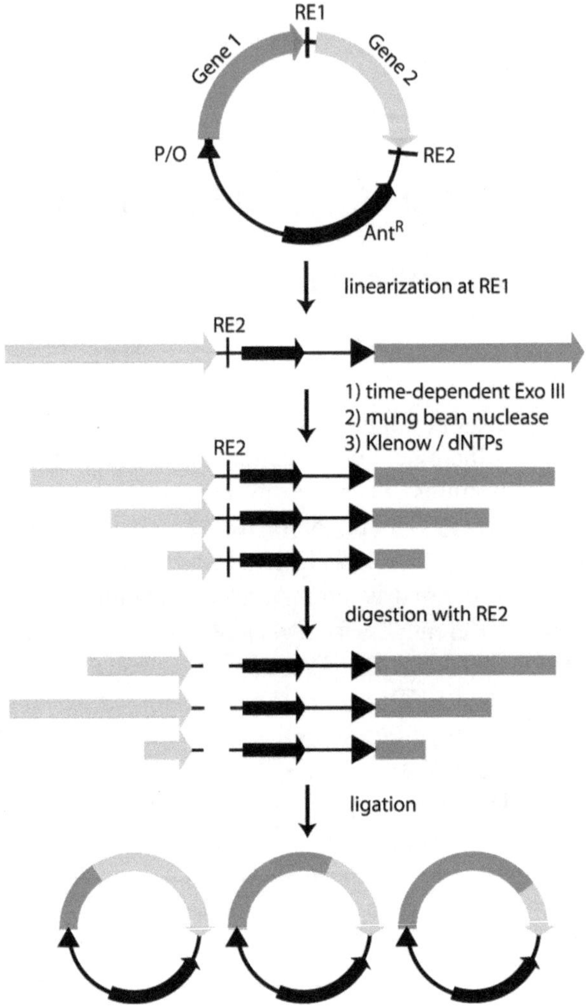

Fig. 1. Schematic of the creation of ITCHY libraries using time-dependent truncation. RE1 and RE2, unique restriction endonuclease sites; P/O, promoter; Ant^R, antibiotic resistance gene.

Each method has advantages and disadvantages. Time dependent truncation requires multiple time-point sampling to achieve a comprehensive library. Truncation using αS-dNTPs is less labor-intensive and offers the possibility of incorporating random mutagenesis in the PCR amplification. A theoretical treatment of both methodologies *(7)* predicts that time-dependent truncation offers higher control over the range of truncation and a higher probability of parental length fusions in the library. However, truncations using αS-dNTPs are predicted to inherently produce a uniform distribution of parental length fusions across the gene regardless of experimental conditions whereas time-dependent truncation only produces such a uniform distribution under optimal experimental conditions.

2. Materials

1. QIAquick PCR and gel purification kits (Qiagen; Valentia, CA) including buffers PB and EB.
2. Restriction endonucleases (with reaction buffers).
3. 100X BSA: 10 mg/mL.
4. NaCl.
5. Exonuclease III (Exo III).
6. 10X Exo III buffer: 660 m*M* Tris-HCl, pH 8.0, 6.6 m*M* $MgCl_2$.
7. Mung Bean Nuclease.
8. 10X Mung bean buffer: 500 m*M* sodium acetate, pH 5.0, 300 m*M* NaCl, 10 m*M* $ZnSO_4$.
9. Klenow fragment DNA polymerase.
10. Klenow mix: 20 m*M* Tris-HCl, pH 8.0, 100 m*M* $MgCl_2$, 0.25 U/μL units Klenow.
11. 10X Klenow buffer: 100 m*M* Tris-HCl, pH 7.5, 50 m*M* $MgCl_2$. Store at room temperature.
12. Deoxynucleotide triphosphates (dNTPs): 2.5 m*M* per dNTP stock solution. Aliquots are stored at –20°C.
13. 100% Ethanol stored at –20°C.
14. 70% Ethanol (v/v) stored at –20°C.
15. Ammonium acetate: 7.5 *M* stored at 4°C.
16. T4 DNA ligase.
17. 10X T4 DNA ligase buffer: 500 m*M* Tris-HCl, pH 7.5, 100 m*M* $MgCl_2$, 100 m*M* dithiothreitol, 10 m*M* ATP, 250 μg/mL bovine serum albumin.
18. Agarose gels: SeaKem GTG agarose (BioWhittaker; Rockland, ME) in 1X TAE buffer (40 m*M* Tris-acetate, 1 m*M* EDTA) and 0.5 μg/mL ethidium bromide.
19. Oligonucleotide primers: 100 μ*M* stock solutions.
20. *Taq* DNA polymerase.
21. 10X *Taq* DNA polymerase buffer: 500 m*M* KCl, 100 m*M* Tris-HCl, pH 9.0, 1.0% Triton X-100.
22. α-Phosphorothioate dNTPs: 100 m*M* stock (Promega; Madison, WI) stored at –20°C.
23. Polyethyleneglycol 45% (w/w): Mix 45 g of polyethylene glycol (MW 6000) and 55 g of water. Autoclave at 121°C for 20 min and let cool to room temperature.
24. Electrocompetent *E. coli* strain DH5α-E (Invitrogen; Carlsbad, CA) stored at –80°C.
25. Luria Bertans Broth (LB) medium: Mix (per liter) 5 g yeast extract, 10 g tryptone, and 10 g NaCl. Autoclave at 121°C for 20 min and let cool to room temperature.

26. LB-agar medium: 5 g yeast extract, 10 g tryptone, 10 g NaCl, and 15 g agar. Autoclave at 121°C for 20 min and let cool to 50°C prior to addition of antibiotic selection marker. Plates can be stored at 4°C for up to 4 wk.
27. SOC medium: Mix (per liter) 5 g yeast extract, 20 g tryptone, 0.5 g NaCl, 2.5 m*M* KCl. Autoclave at 121°C for 20 min and let cool to room temperature. Autoclave separately 1 *M* $MgCl_2$, 1 *M* $MgSO_4$, and 20% (w/v) glucose. Add 10 m*M* $MgCl_2$, 20 m*M* $MgSO_4$, and 20 m*M* glucose to the media. Store at room temperature.

3. Methods

The methods described below outline 1) the construction of ITCHY libraries using time-dependent truncation, 2) the construction of ITCHY libraries using nucleotide analogs and 3) characterization of the ITCHY libraries. Throughout these methods, we utilize Qiagen's QIAQuick PCR purification and gel extraction kits to purify and desalt DNA solutions. DNA purification kits from other manufacturers presumably work equally well but they have not been tested in our laboratory.

3.1. Preparation of Incremental Truncation Libraries Using Time-Dependent Truncation

1. Construct a vector as shown in **Fig. 1**. The two parental gene sequences are cloned in series, using unique restriction sites in the vector (*see* **Note 1**).
2. Linearize 10 µg of non-nicked plasmid DNA (*see* **Note 2**) by digestion with a restriction enzyme RE1 (*see* **Note 3**) that recognizes a unique cleavage site between the two parental genes (*see* **Fig. 1**). Purify the reaction mixture using Qiagen's QIAquick PCR purification kit, eluting with 100 µL of buffer EB provided by the QIAquick kit. Determine the DNA concentration by absorbance at 260 nm using the relationship that 50 µg/mL gives an A_{260} of 1.0.
3. Equilibrate 300 µL of PB buffer (provided in Qiagen's QIAquick PCR purification kit) on ice in a 1.5-mL tube (tube A).
4. To a second 0.65-mL tube (tube B), add 2 µg of DNA from *step 2*, 6 µL of 10X Exo III buffer, and NaCl to the desired concentration (*see* **Note 4**), water to 60 µL and equilibrate at 22°C. There are useful controls that can be incorporated at this step (*see* **Note 5**).
5. At time = 0, add 100 units of Exo III per µg DNA to tube B and mix immediately and thoroughly.
6. At regular intervals (generally 20–30 s) remove small samples (~0.5–1 µL) and add to tube A (*see* **Note 6**). Mix tube A well. Note that all time points are removed to tube A, which is kept on ice. The rate of Exo III is very temperature dependent; thus, it is preferable to leave tube B open during the sampling to avoid warming the tube by repeated handling.
7. Purify the DNA using Qiagen's QIAquick PCR purification kit following the manufacturer's directions. Elute the DNA from the column with 47 µL of Buffer EB.
8. Add 5 µL of mung bean buffer (10X) and mix thoroughly.

9. Add 3 U of mung bean nuclease, mix thoroughly and incubate at 30°C for 30 min (*see* **Note 7**).
10. Add 250 μL of QIAquick buffer PB and purify the DNA using Qiagen's QIAquick PCR purification kit following the manufacturer's directions. Elute truncated DNA from QIAquick column with 82 μL of Buffer EB.
11. Equilibrate tube at 37°C.
12. Add 0.5 μL of dNTP stock solution and 10 μL of Klenow mix, mix thoroughly and incubate at 37°C for 5 min.
13. Inactivate Klenow by incubating at 72°C for 20 min.
14. Add 10 μL of the appropriate 10X restriction enzyme buffer for restriction enzyme RE2, 1 μL BSA (100X) and an appropriate number of units of restriction enzyme RE2 (*see* **Note 8**) and incubate at the appropriate temperature for 2 h. Heat-inactivate RE2 under the conditions recommended by the manufacturer.
15. Concentrate by ethanol precipitation: add 50 μL of ammonium acetate followed by 300 μL of 100% ethanol (*see* **Note 9**). Incubate on ice for 30 min. Centrifuge 10 min at 12,000*g*. Wash the pellet with 70% ethanol and centrifuge 2 min at 12,000*g*. Remove liquid by pipet, spin briefly again and remove all traces of liquid. Air-dry the pellet 10 min and resuspend in 17 μL water.
16. Add 2 μL of T4 DNA ligase buffer (10X) and 6 Weiss units of T4 DNA ligase and incubate at 15°C ≥12 h.
17. Add 30 μL of water and ethanol precipitate as in **step 15** into 20 μL of water. The DNA is now ready to be transformed into the desired host.

3.2. Preparation of Incremental Truncation Libraries Using Nucleotide Analogs

1. Construct a template vector as shown in **Fig. 2**. The two parental gene sequences are cloned in series, using unique restriction sites in the vector (*see* **Note 10**).
2. Linearize the template vector (~ 3 μg) by digestion with a restriction enzyme that recognizes a unique cleavage site between the two parental genes (RE in **Fig. 2**). Purify the reaction mixture by agarose gel electrophoresis (*see* **Note 11**). Pour a 1% agarose gel with sufficiently large well to load entire digest. Load digest in well after mixing it with loading dye and run gel. Visualize the bands of DNA under UV light and excise product band that corresponds in size to the linearized vector with a razor blade. Recover DNA from gel matrix, using Qiagen's QIAquick gel purification kit.
3. Set up 50 μL PCR reactions containing 1X *Taq* DNA polymerase buffer, 1.25 m*M* $MgCl_2$, 0.5–1 μ*M* primer A and B (*see* **Fig. 2**), ~10 ng linearized template vector, 5 units *Taq* DNA polymerase (*see* **Note 12**), and 200 μ*M* dNTP (control experiment), 180 μ*M* dNTP / 20 μ*M* α-phosphorothioate dNTP, or 175 μ*M* dNTP/25 μ*M* α-phosphorothioate dNTP. Amplify template over 30 cycles of PCR (94°C for 30 sec, T_A for 30 sec, 72°C for 1.2 min/kb) (*see* **Note 13**). For all subsequent steps, the control experiment (dNTP only), as well as the individual dNTP/αS-dNTP experiments are treated in parallel.

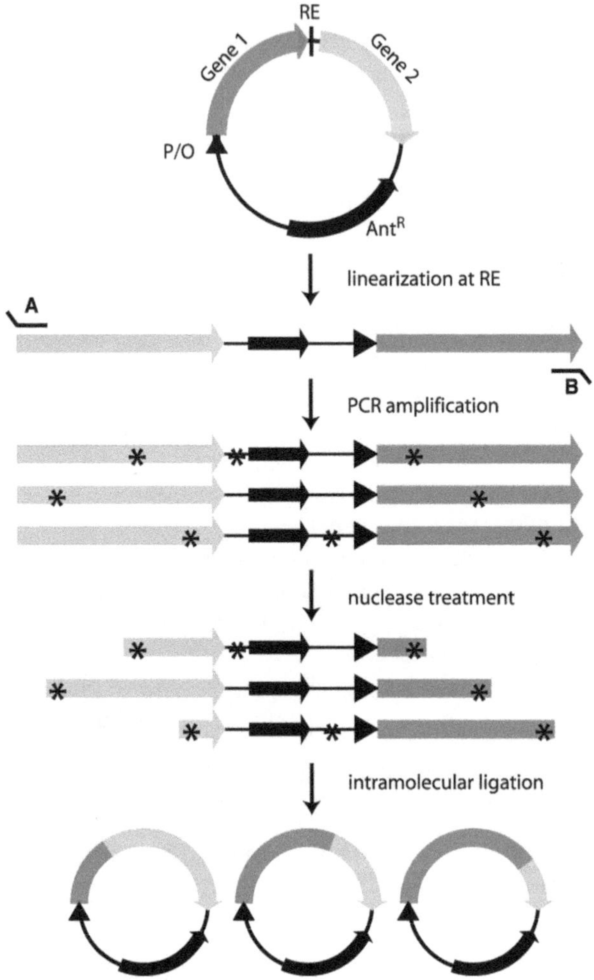

Fig. 2. Schematic of the creation of ITCHY libraries using nucleotide analogs. RE, unique restriction endonuclease sites; P/O, promoter; Ant^R, antibiotic resistance gene; *, site of nucleotide analog incorporation. **A** and **B**, site-specific amplification primers.

4. Check quality of the PCR products by running 1–2 μL of reaction mixture on an agarose gel next to a suitable size marker. Purify the remaining reactions with Qiagen's QIAquick PCR purification kit (*see* **Note 14**).
5. Spectrophotometrically determine the final DNA concentration of all samples (*see* **Note 15**).
6. Dilute PCR product to approx 0.1 μg/μL of DNA in 1X Exo III buffer, using water and 10X Exo III buffer. The solution is centrifuged (at 5000*g* at room temperature for 1 min).

7. Add Exo III (120 units per μg of DNA) and digest the reaction mixture at 37°C for 30 min (*see* **Note 16**).
8. Quench reaction by addition of 5 volumes of PB buffer (provided in Qiagen's QIAquick PCR purification kit) and recover the DNA with a spin column according to the manufacturer's protocol (*see* **Note 17**).
9. Dilute DNA solution to approx 0.1 μg/μL of DNA in 1X mung bean nuclease buffer, using water and 10X mung bean nuclease buffer.
10. Add 2.5 units of mung bean nuclease per μg of DNA. Briefly centrifuge the samples and incubate them at 30°C for 30 min.
11. Quench reaction by addition of 5 volumes of PB buffer (provided in Qiagen's QIAquick PCR purification kit) and recover DNA with spin columns according to the manufacturer's protocol (*see* **Note 17**).
12. Mix DNA solution to approx 0.1 μg/μL of DNA in 1X Klenow buffer, using water and 10X Klenow buffer, spin briefly, and incubate 15 min at 37°C (*see* **Note 18**).
13. Add 1 unit of Klenow fragment DNA polymerase per μg of DNA and continue incubation for 5 min at 37°C (*see* **Note 18**).
14. Add dNTP stock solution to a final concentration of 100 μ*M* and continue incubation at 37°C for 10 min (*see* **Note 18**).
15. Quench reaction by heat-denaturation for 20 min at 75°C. After equilibration at room temperature, quality control can be performed by running an aliquot of the reaction mixture on an agarose gel next to a suitable size marker (*see* **Note 19**).
16. Ligate the remaining truncation product in 400 μL, containing 1X T4 ligase buffer (provided with enzyme), 4.5% PEG-6000, and 5 Weiss-units of T4 DNA ligase per μg of DNA, by incubating the mixture at 16°C overnight (*see* **Note 20**).
17. Quench reaction by addition of 3 volumes of QG buffer (provided in Qiagen's QIAquick gel extraction kit) and recover DNA with spin columns according to the manufacturer's protocol. Elute DNA in a final volume of 50 μL EB buffer (provided in the QIAquick kit). The DNA is now ready to be transformed into the desired host.

3.3. Characterization of Incremental Truncation Libraries

1. Transform aliquots of library into electrocompetent *E. coli* (*see* **Note 21**). (a) Thaw one 50 μL-aliquot of competent cells on ice for each sample. (b) Add up to 20 ng of sample DNA per aliquot of competent cells, then transfer the cell suspension into an ice-cold electroporation cuvet and store the cuvettes on ice for 1 min. (c) Load the cuvet into the electroporator and apply an electrical pulse (for optimal settings of electroporator, contact manufacturer's manual). (d) Remove cuvet from electroporator and immediately add 1 mL of SOC medium. Incubate the cell suspension at 37°C for 1 h. (e) Plate various volumes of cell suspension (5–100 μL) onto LB-agar plates containing the appropriate antibiotic selection marker and incubate the plates at 37°C overnight.
2. Assess the apparent library size by counting the colony numbers on the culture plates, considering the applied dilutions.

3. Determine the fraction of the library containing inserts in the desired range by colony-PCR (*see* **Note 22**) as follows: Pick 25–50 colonies with sterile toothpicks and suspend cells in 10 µL PCR reaction mixtures (*see* **Note 23**). Initiate amplification with a five-min denaturing/cell lysis step at 95°C, followed by the regular cycling protocol. Mix reaction product with gel loading dye and run an aliquot on an agarose gel next to an appropriate size marker. Count the fraction of samples that generated a PCR product within the desired size range (*see* **Note 24**).
4. Analyze the data from the colony-PCR with respect to actual library size and the fragment size distribution of the library. (*see* **Note 25**) The position of the precise fusion point in selected sequences can be determined by DNA sequencing.

4. Notes

1. There are no restrictions on the vector that are particular to the construction of ITCHY libraries except for the uniqueness of the restriction enzyme site between the cloned genes and that it's digestion produces 3' ends that are susceptible to Exo III digestion (e.g., has blunt or 3' recessed end).
2. Generally we use plasmid DNA that is ~90% or greater supercoiled. We routinely isolate plasmid DNA of such quality from *E. coli* strain DH5α using commercial plasmid-prep kits. We have found that the fraction of nicked molecules can depend on growth conditions ***(8)***.

   Should production of sufficiently pure non-nicked DNA prove difficult, methods to purify non-nicked from nicked plasmid DNA include CsCl-ethidium bromide gradients ***(9)***, acid-phenol extraction ***(10)***, or removal of the nicked DNA by enzymatic digestion ***(11)***. Alternatively, treatment of nicked DNA with T4 DNA ligase should presumably repair the nicks.
3. Another important factor to consider in preparing non-nicked DNA for incremental truncation is the restriction enzyme digestion to linearize the DNA in preparation for truncation. Restriction enzymes from suppliers may have nuclease contamination. In addition, restriction enzymes may have single-stranded nicking activity at high enzyme to DNA ratio. For this reason, we recommend digesting the DNA with the minimum amount of restriction enzyme necessary to fully digest the DNA and avoiding conditions contributing to star activity (relaxed or altered specificity) in restriction enzymes (*see* manufacturer's product specification). On the other hand, it should be emphasized that incomplete digestion at this step could result in a significant fraction of the final library being untruncated. Gel purification of the digested DNA can guard against this (*see* **step 2** in **Subheading 3.2.**).
4. We found the rate of truncation to vary with NaCl at 22°C by the following equation: rate (bp/min)= $47.9 \times 10^{(-0.00644 \times N)}$ where N = concentration of NaCl in m*M* (0–150 m*M*). Using this equation, the rate of Exo III digestion is ~10 bases/min at 22°C in the presence of 100 m*M* salt. This rate expression is valid for a DNA concentration of 33.3 ng/µL and 100 units of Exo III per µg DNA. The rate of Exo III in the presence of NaCl at higher temperatures has been determined elsewhere ***(12)***.

5. It is useful to include internal controls for testing fidelity and rate of truncation. If using these controls, increase the amount of DNA to 6 μg , the amount of buffer to 18 μL, and the total volume to 180 μL while keeping the desired NaCl concentration. Remove a 60 μL aliquot (a no truncation control) to a tube of 300 μL PB buffer before adding the Exo III. At some point during the truncation (a final time-point works well), remove one 60-μL time point to a tube of 300 μL PB buffer. Process both these tubes in parallel with the truncation library through the mung bean nuclease step. At this point analyze the DNA in these control tubes by agarose gel electrophoresis to determine the experimental truncation rate. To make size determination more accurate, or in cases where the length of truncation is small in comparison to the size of the plasmid, it may be advantageous to digest this DNA with a restriction enzyme prior to electrophoresis.
6. The sample size depends on the maximum truncation length, the truncation rate and the frequency of sampling. For example, to truncate 300 bp, one could digest at 100 m*M* salt (Exo III digestion rate ~ 10 bp/min) and take 1 μL samples every 30 sec for 30 min. Exo III digestion produces a Gaussian distribution of truncation lengths with a standard deviation of 0.075–0.2 times the total truncation length ***(12)***. Thus, one can take successive samples whose average truncation length differs by more than one base and still expect a relatively even distribution of truncation lengths ***(7)***.
7. Occasionally we have had to optimize this step by performing test mung bean nuclease digestions. We use a single truncation time point for the test (i.e., digest 12 μg of DNA with Exo III for the time necessary to digest 100 bp). The single time point is then divided into six tubes of 47 μL each (i.e., 2 μg DNA in each tube, as in the normal truncation protocol) to which is added 5 μL of mung bean buffer (10X) and mung bean nuclease in different amounts (try 12, 6, 3, 1.5, and 0.75, and 0 units). A portion of the DNA is then run on a 0.7% agarose gel. DNA not digested with mung bean nuclease will run in a somewhat diffuse band larger than the expected size owing to the large, single-stranded overhangs. As more mung bean nuclease is added, the DNA will begin to smear between the size obtained without mung bean nuclease treatment and the expected size. At the optimum amount of mung bean nuclease, the DNA will run as a relatively focused band at the expected size. When too much mung bean nuclease is added, the DNA will begin to smear to much smaller sizes as mung bean nuclease begins to make double stranded breaks.
8. It is very important that digestion is complete so as to avoid a significant fraction of the final library being a coordinated truncation library (i.e., both genes truncated approx the same amount).
9. In our experience, ammonium acetate is preferable over sodium chloride or sodium acetate. To help visualize the precipitated pellet in subsequent steps, we usually add 2 μL of pellet paint (Novagen; Madison, WI) after the ammonium acetate and before the 100% ethanol.
10. No particular restrictions apply to what plasmid(s) can or cannot be used as template vector as long as sufficient unique restriction sites are present. However,

minimizing the size of the vector is advantageous for the subsequent PCR amplification step, reducing the required extension time and minimizing the accumulation of point mutations. Template vectors (including parental genes) of up to 5000 basepairs have successfully been used.

11. Gel purification of the digestion product is very important to prevent contamination of the subsequent steps in the protocol and the final library with uncut vector.
12. The preferred polymerase is *Taq* DNA polymerase since it has the lowest error-rate of exonuclease-deficient polymerases. You can use any heat-stable DNA polymerase provided that it is exonuclease-deficient. Upon completion of DNA synthesis, polymerases with exonuclease activity engage in a process known as idling: the continuous removal and resynthesis of the 3'-ends. None of the 3'-5' exonuclease activities of Klenow, T4, *Vent*, and *Pfu* DNA polymerase is capable of hydrolyzing the phosphorothioate linkage. Thus, this idling leads to the unwanted accumulation of phosphorothioate containing nucleotides at the 3'-ends of the resynthesized strands, biasing the resulting library towards full-length fragment sizes.
13. The best ratio of dNTPs to α-phosphorothioate dNTPs depends on the size of the truncation region. The two listed mixtures work well for truncations between 100–1000 basepairs. Larger quantities of PCR product for the following steps may be desired. We recommend to run two or three 50 µL reactions of each dNTP / α-phosphorothioate dNTP mixture in parallel, rather than scaling up the reaction itself. The optimal annealing temperature for primer A and B must be determined prior to the amplification with α-phosphorothioate dNTPs. We recommend primers of 20–25 bp length with T_A of 55–60°C.
14. Each 50 µL PCR reaction generates 1–10 µg of product. The quantity and quality of the product bands of the control reaction and the various dNTP/αS-dNTP ratios should be more or less the same. We usually combine a maximum of two reactions per spin column to minimize product losses.
15. We use the following rule for concentration measurements: 1.0 A_{260} = 50 µg DNA. As a guideline, our experimental yields range from 5–20 µg of PCR product in 50–100 µL of elution buffer (10 m*M* Tris-HCl, pH 8, provided by Qiagen as part of the QIAquick kits).
16. It is very important to carefully but thoroughly mix the reaction upon addition of Exo III. We obtained good, consistent results by using a pipettor for 20–30 s.
17. The final step of the QIAquick purification protocol consists of eluting the DNA with a minimum of 30 µL EB-buffer. We usually adjust this volume to directly obtain DNA concentrations of approx 1 µg of DNA per 10 µL solution. Our calculations are based on the initial absorbance measurements in **step 5** and **not** on measurements after nuclease treatment.
18. During the blunt-ending as described in **steps 10–12**, each consecutive addition of reagent involves a careful mixing of the reaction solution, using the pipettor. Attention should be paid to splashing parts of the reaction mixture. In those cases, a brief spinning of the sample is recommended.
19. We recommend loading 1–4 µg of DNA (again based on the original absorbance measurements) to ensure a sufficient amount of DNA for visualization. A typical

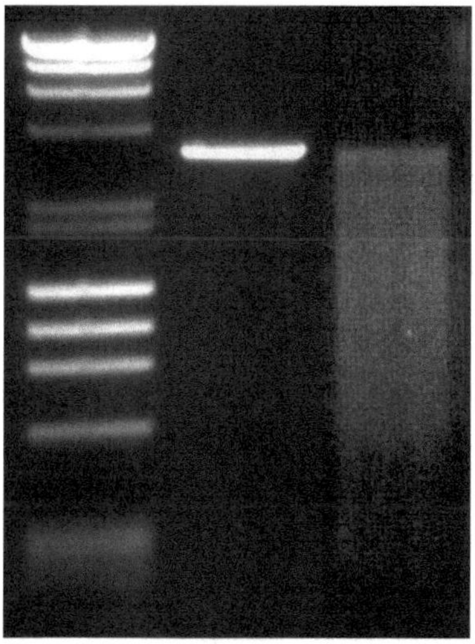

Fig. 3. Exo III treatment of PCR-amplified linearized plasmid containing nucleotide analogs. Size marker is a 1:1 mixture of Lamda DNA (*Hin*dIII digest) and ϕX174 DNA (*Hae*III digest). The center shows the product band after the PCR amplification in the presence of nucleoside analogs. On the right, the same sample after nuclease treatment is a smear of truncated material, reaching from the size of the original construct to very small fragments of a few hundred nucleotides in length.

example for an incremental truncation library at this stage is shown in **Fig. 3**. Note that the control experiment with only dNTPs should have no visible band at all.

20. Crucial to the successful intramolecular ligation is to maintain an overall DNA concentration of <3 ng/μL *(8)*. Adjust volumes of the ligation reaction for DNA concentration to fall below this threshold.
21. We have had consistently good results with strain DH5α-E. Transformation by electroporation is generally more efficient, generating larger libraries than chemical transformation, however either method works for transforming incremental truncation libraries.
22. Although the colony PCR method is quick and easy, some samples fail to amplify a band even though they have truncations in the desired range. A more definitive, but time consuming, method to characterize the size distribution of the library is to grow inoculum of randomly selected colonies, miniprep the plasmid DNA, digest the DNA with the appropriate restriction enzymes and analyze the distribution by agarose gel electrophoresis.
23. PCR reaction mixture consists of 1X *Taq* polymerase buffer, 1.25 m*M* $MgCl_2$, 200 μ*M* dNTPs, and 1 μ*M* gene-specific primers.

24. In our experience, approx 50% of the colony-PCR reactions generate a PCR product, ranging in size from completely untruncated material to fragments only a few dozen bases in length. The ratio of insert-carrying colonies can be improved by size selection of the incremental truncation library on an agarose gel, subsequent to the blunt-ending step.
25. As a rule-of-thumb, in order for the library to have the vast majority of possible crossovers, the number of transformants with truncations in the correct range should be approx 5–10 times the theoretical degeneracy of the library. The theoretical degeneracy of an ITCHY library is the product of the maximum number bases truncated in each gene. The following more rigorous treatment of library completeness derives from a treatment of genomic libraries *(13)*. The probability P_C that a library is complete can be calculated as follows. We will treat the ITCHY library as ideal in that only truncations in the desired range are created and all members are equally represented. Deviations from ideality are discussed below. Let T_{max1} and T_{max2} be the maximum desired truncations (in basepairs) of genes 1 and 2 respectively. Then the degeneracy *D* of the library (the number of possible variants) is the product of T_{max1} and T_{max2}. The frequency, *f*, of a particular fusion in this library is the inverse of the degeneracy (1/*D*). The probability of not picking a particular sequence if one picks a random library member is 1-*f*. Thus, given a number of transformants *N*, the probability P_i that sequence *i* is in the library is given by **Equation 1**.

$$P_i = 1 - [1 - f]^N \qquad \text{[Eq. 1]}$$

Using **Equation 1**, it can be shown that one needs to have the number of transformants be 4.6-fold more than the degeneracy *D* in order to have a 99% probability of having a particular sequence in the library. This factor is essentially independent of *D* for $D > 100$. The probability that a library is complete, P_C, is the product of all P_i's for the entire degeneracy given by **Equation 2**.

$$P_C = [1 - (1 - f)^N]^D \qquad \text{[Eq. 2]}$$

Using **Equation 2**, it can be shown that the factor $F_{0.99}$ by which number of transformants *N* must exceed the degeneracy *D* in order that there is a 99% probability that the library is complete ranges from about 10 to 20 for typical library sizes and can be given by **Equation 3**.

$$F_{0.99} = 4.4049 + 2.3348\log(D) \qquad \text{[Eq. 3]}$$

Equations 1–3 are not particular to ITCHY libraries and can be applied to any library by appropriately calculating its degeneracy. Finally, the above discussion of probabilities should be tempered by three factors. First, libraries are never ideal, and thus the true *f* is the product of the inverse of the degeneracy and the fraction of the library that has truncations in the desired range. Second, the above discussion assumes that the all members of the library are equally represented. Modeling indicates that even under the best of experimental conditions, this will not be true for either method, particularly truncations using nucleotide analogs *(6)*. Third, while a complete library is a good goal to have, libraries do not have to be complete to be useful.

Acknowledgments

We thank Stephen J. Benkovic for his support and guidance. We also thank Dave Paschon for optimizing the mung bean nuclease step in the time-dependent truncation protocol.

References

1. Ostermeier, M., Nixon, A. E., Shim, J. H., and Benkovic, S. J. (1999) Combinatorial protein engineering by incremental truncation. *Proc. Natl. Acad. Sci. USA* **96,** 3562–3567.
2. Lutz, S., Ostermeier, M., and Benkovic, S. J. (2001) Rapid generation of incremental truncation libraries for protein engineering using α-phosphothioate nucleotides. *Nucleic Acids Res.* **29,** e16.
3. Ostermeier, M., Shim, J. H., and Benkovic, S. J. (1999) A combinatorial approach to hybrid enzymes independent of DNA homology. *Nat. Biotechnol.* **17(12),** 1205–1209.
4. Sieber, V., Martinez, C. A., and Arnold, F. H. (2001) Libraries of hybrid proteins from distantly related sequences. *Nat. Biotechnol.* **19(5),** 456–60.
5. Lutz, S., Ostermeier, M., Moore, G. L., Maranas, C. D., and Benkovic, S. J. (2001) Creating multiple-crossover DNA libraries independent of sequence identity. *Proc. Natl. Acad. Aci. USA* **98,** 11,248–11,253.
6. Putney, S. D., Benkovic, S. J., and Schimmel, P. R. (1981) A DNA fragment with an alpha-phosphorothioate nucleotide at one end is asymmetrically blocked from digestion by exonuclease III and can be replicated *in vivo*. *Proc. Natl. Acad. Aci. USA* **78,** 7350–7354.
7. Ostermeier, M. (2003) Theoretical distribution of truncation lengths in incremental truncation libraries. *Biotechnol. Bioeng*., in press.
8. Ostermeier, M., Lutz, S., and Benkovic, S.J. (2002) Generation of protein fragment libraries by incremental truncation, in *Protein-Protein Interactions: A Molecular Cloning Manual*, (Golemis, E. A., ed.), Cold Spring Harbor Laboratory Press, Cold Spring Harbor, NY.
9. Sambrook, J., Fritsch, E. F., and Maniatis, T. (1989) *Molecular Cloning: A Laboratory Manual, 2nd ed*, Cold Spring Harbor Laboratory Press, Cold Spring Harbor, NY.
10. Zasloff, M., Ginder, G. D., and Felsenfeld, G. (1978) A new method for the purification and identification of covalently closed circular DNA molecules. *Nucleic Acids Res.* **5,** 1139–1152.
11. Gaubatz, J. W. and Flores, S. C. (1990) Purification of eucaryotic extrachromosomal circular DNAs using exonuclease III. *Anal. Biochem.* **184,** 305–310.
12. Hoheisel, J. D. (1993) On the activities of *Escherichia coli* exonuclease III. *Anal. Biochem.* **209,** 238–246.
13. Zilsel, J., Ma, P. H., and Beatty, J. T. (1992) Derivation of a mathematical expression useful for the construction of complete genomic libraries. *Gene* **120,** 89–92.

17

Preparation of SCRATCHY Hybrid Protein Libraries

Size- and In-Frame Selection of Nucleic Acid Sequences

Stefan Lutz and Marc Ostermeier

1. Introduction

SCRATCHY is a combination of the incremental truncation for the creation of hybrid enzymes (ITCHY) technology *(1)* and DNA shuffling *(2)*. It generates combinatorial libraries of hybrid proteins consisting of multiple fragments from two or more parental DNA sequences with no restriction to DNA sequence identity between the original sequences *(3)*. Such multi-crossover hybrids can be of interest to the studies of fundamental questions of protein evolution and folding, as well as to the tailoring of enzymes for therapeutic and industrial applications.

The experimental implementation of SCRATCHY consists of two successive steps, an initial creation of an ITCHY library, followed by a homologous recombination procedure such as DNA shuffling (*see* **Fig. 1**). In the process, the ITCHY library serves as an artificial family of hybrid sequences that, upon fragmentation, provides a variety of crossover-carrying templates. During the random reassembly step, these templates can anneal with one another, leading to the introduction of two or more crossovers per shuffled sequence independent of the sequence homology in any particular region.

Two experimentally critical steps of SCRATCHY lie at the interface of the ITCHY and DNA shuffling protocols. First, a significant fraction of the incremental truncation library members contain large sequence insertions and deletions. Presumably, these deviations from the parental-size gene sequences are likely detrimental to the library and should be removed by size selection. Second, the blunt-end fusion of the *N*- and C-terminal fragment of the truncation library results in the disruption of the correct nucleotide-codon reading frame

From: *Methods in Molecular Biology, vol. 231: Directed Evolution Library Creation: Methods and Protocols*
Edited by: F. H. Arnold and G. Georgiou © Humana Press Inc., Totowa, NJ

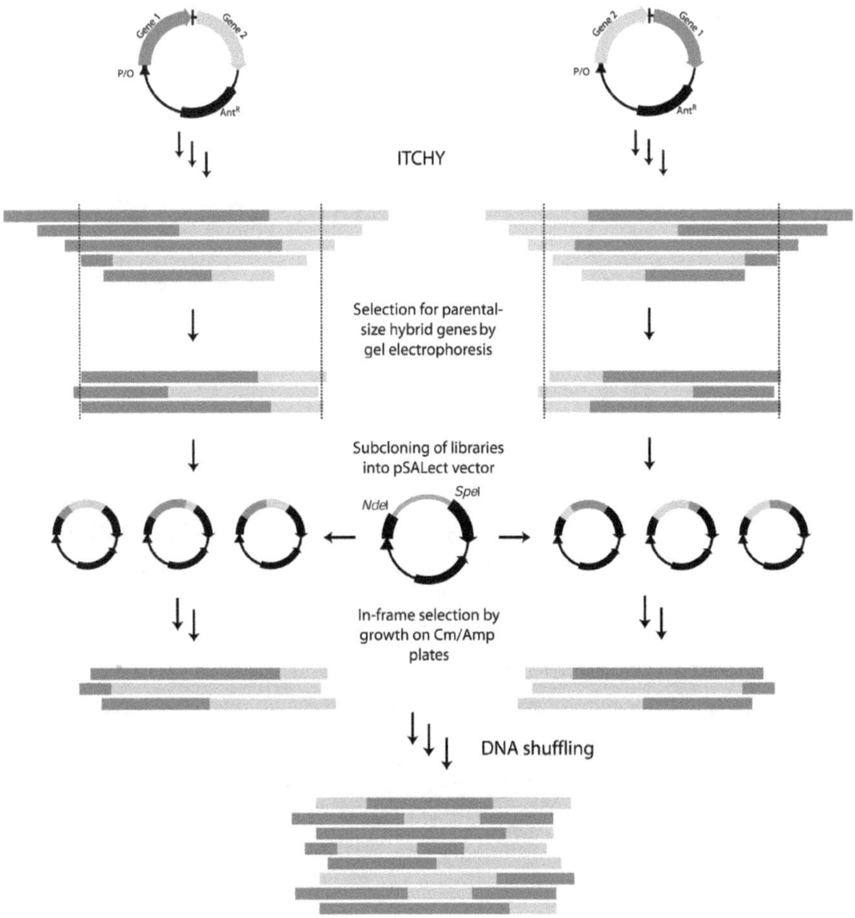

Fig. 1. Schematic view of the SCRATCHY protocol. The method requires two complementary vectors, carrying the genes A and B in alternating order. Following the generation of the ITCHY library, the linearized hybrid genes are selected for parental-size hybrid DNA constructs. After subcloning these DNA fragments into the *Nde*I and *Spe*I sites of pSALect, sequences with the correct reading frame result in the expression of a trifunctional fusion protein which renders the host cells resistant to ampicillin. The plasmid DNA from colonies grown under these conditions is recovered and can be used as starting material for DNA shuffling.

at the crossover in two-thirds of the library members. While not crucial for the preparation of ITCHY libraries, the presence of such out-of-frame fusions will rapidly diminish the fraction of hybrids with the overall correct reading frame upon DNA shuffling. For example, the probability of a hybrid with two fusion

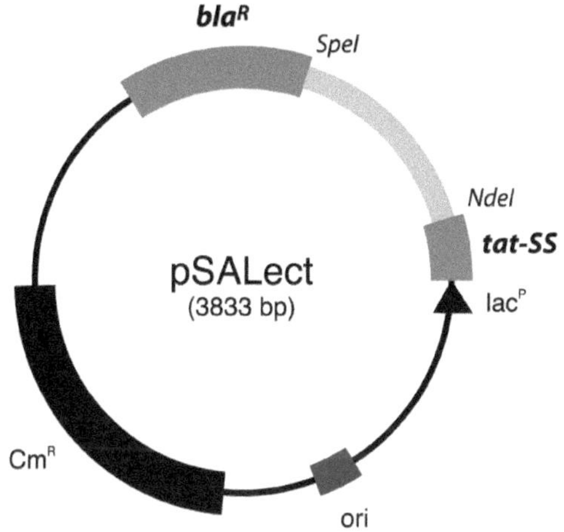

Fig. 2. The pSALect vector is a pBC-SK+ based plasmid for in-frame selection of DNA sequences. The gene of interest is cloned through *Nde*I and *Spe*I sites that are flanked *N*-terminally by the Tat-signal sequence (tat-SS) and C-terminally by the gene encoding for β-lactamase (bla^R) without its natural signal sequence. The expression of the fusion construct is under the control of a *lac*-promoter (lac^P). As a general selection marker, the plasmid carries the gene for chloramphenicol-acetyl-transferase (Cm^R).

points maintaining the correct reading frame is only 1/9 (1/3 × 1/3) while for a hybrid with three crossovers, the chances drop to 1/27 (1/3 × 1/3 × 1/3). Addressing these problems, we have developed and implemented a two-step post-ITCHY protocol to remove fragments with large insertions and deletions and out-of-frame constructs.

This chapter describes the two-step protocol for DNA fragment size- and reading frame-selection of ITCHY libraries prior to DNA shuffling. Initially, the linearized version of the plasmid DNA carrying the ITCHY library is separated by agarose gel electrophoresis. The desired size-range of fragments is excised and recovered. Next, the size-selected sublibrary is cloned into pSALect and in-frame selection is performed. The pSALect vector is specifically designed to allow for the isolation of nucleic acid sequences with the correct reading frame *(4)*. The target sequence is cloned between an *N*-terminal Tat signal sequence and a C-terminal lactamase gene (*see* **Fig. 2**). Upon expression, the target peptide acts as a linker between the two selection markers. For in-frame fusions, this tri-functional fusion facilitates the successful export of the lactamase into the periplasm, rendering the host cell resistant to antibiotics in the growth media. This strategy eliminates the false positives, presumably arising from internal translational initiation sites, observed in in-frame selection strategies using *N*-terminal fusions to GFP ***(5)***, chloramphenicol-acetyl-transferase ***(6)*** and aminoglycoside phosphotransferase APH(3')-IIa (Ostermeier, M., Lutz, S., and Benkovic, S.J., unpublished results), since functional expression of the lactamase requires export to the periplasm. After in-frame selection, the plasmid DNA from the selected candidates is recovered and can be used for DNA shuffling.

2. Materials

1. QIAquick PCR and gel purification kits (Qiagen; Valentia, CA) including buffers PB and EB.
2. Restriction endonucleases (with reaction buffers).
3. Glycerol.
4. Chloramphenicol.
5. Ampicillin.
6. T4 DNA ligase.
7. 10X T4 DNA ligase buffer: 500 m*M* Tris-HCl, pH 7.5, 100 m*M* $MgCl_2$, 100 m*M* dithiothreitol, 10 mM ATP, 250 µg/mL bovine serum albumin (BSA).
8. Polyethyleneglycol 45% (w/w): Mix 45 g of polyethylene glycol (MW 6000) and 55 g of water. Autoclave at 121°C for 20 min and let cool to room temperature.
9. Shrimp alkaline phosphatase: USB Corporation (Cleveland, OH).
10. Agarose gels: NuSieve GTG and SeaKem GTG agarose (both from BioWhittaker; Rockland, ME) in 1X TAE buffer (40 m*M* Tris-acetate, 1 m*M* EDTA) and 0.5 µg/mL ethidium bromide.
11. 100 bp-DNA Step Ladder: Promega (Madison, WI) stored at –20°C.
12. Electrocompetent *E. coli* strain DH5α-E (Invitrogen; Carlsbad, CA) stored at –80°C.
13. Luria Bertani Broth (LB) medium: Mix (per liter) 5 g yeast extract, 10 g tryptone, and 10 g NaCl. Autoclave at 121°C for 20 min and let cool to room temperature.
14. LB-agar medium: 5 g yeast extract, 10 g tryptone, 10 g NaCl, and 15 g agar. Autoclave at 121°C for 20 min and let cool to 50°C prior to addition of antibiotic selection marker. Plates can be stored at 4°C for up to 4 wk.
15. SOC medium: Mix (per liter) 5 g yeast extract, 20 g tryptone, 0.5 g NaCl, 2.5 m*M* KCl. Autoclave at 121°C for 20 min and let cool to room temperature. Autoclave separately 1 *M* $MgCl_2$, 1 *M* $MgSO_4$, and 20 % (w/v) glucose. Add 10 m*M* $MgCl_2$, 20 m*M* $MgSO_4$, and 20 m*M* glucose to the media. Store at room temperature.
16. pSALect plasmid DNA vector: Stefan Lutz (Emory University, Atlanta, GA)
17. Deoxynucleoside triphosphates (dNTPs): 2.5 m*M* per dNTP stock solution. Aliquots are stored at –20°C.
18. Oligonucleotide primers: 100 µ*M* stock solutions.
19. *Taq* DNA polymerase.
20. 10X *Taq* DNA polymerase buffer: 500 m*M* KCl, 100 m*M* Tris-HCl, pH 9.0, 1.0% Triton X-100.

3. Methods

In the preparation of an incremental truncation library for DNA shuffling, two additional, consecutive steps are highly recommended: a) the removal of truncated fragments with large insertions and deletions and b) the isolation of single-crossover hybrids that preserve the correct reading frame.

The following protocol builds on the chapter on the creation of ITCHY hybrid protein libraries (*see* **Chapter 16**). The starting material used in the procedure originates from an ITCHY experiment using nucleoside analogs to create the incremental truncation library. The protocol starts after the blunt-ending of the library by Klenow fragment DNA polymerase (**Chapter 16**; **Subheading 3.2.**, **step 15**). Throughout the protocol, we utilize Qiagen's QIAQuick PCR purification and gel extraction kits to purify and desalt DNA solutions. DNA purification kits from other manufacturers presumably work equally well but they have not been tested in our laboratory.

3.1. Size Selection of DNA Fragments

1. Purify the heat-denatured reaction mixture of the incremental truncation library by agarose gel electrophoresis: Pour a 1% agarose gel with two wells, sufficiently large to load the entire truncation library (*see* **Note 1**), Load aliquots of the reaction mixture in well after mixing them 1:1 with glycerol and run gel at a low voltage, Visualize the DNA library smear under UV light and excise the gel block that contains DNA product, corresponding to the desired size range (*see* **Note 2**). Recover DNA from gel matrix, using Qiagen's QIAquick gel purification kit (for details see manufacturer's protocol).
2. Ligate the product in 400 µL, containing 1X T4 ligase buffer (provided with enzyme), 4.5% PEG-6000, and 5 Weiss-units of T4 DNA ligase per µg of DNA, by incubating the mixture at 16°C overnight (*see* **Note 3**).
3. Quench reaction by addition of 3 volumes of QG buffer (provided in Qiagen's QIAquick gel extraction kit) and recover DNA with spin columns according to the manufacturer's protocol. Elute DNA in a final volume of 50 µL EB buffer (provided in the QIAquick kit).
4. Transform 2-µL aliquots of the library into electrocompetent *E. coli* (*see* **Note 4**):
   a. Thaw one frozen 50-µL aliquot of competent cells for each sample by storing the vials on ice.
   b. Add up to 20 ng of sample DNA per aliquot of competent cells, then transfer the cell suspension into an ice-cold electroporation cuvette, and store the cuvette on ice for 1 min.
   c. Load the cuvet into the electroporator and apply an electrical pulse (for optimal settings of electroporator, contact manufacturer's manual).
   d. Remove cuvet from electroporator and immediately add 1 mL of SOC medium. Incubate the cell suspension at 37°C for 1 h.

e. Dilute a two-microliter culture sample in 1 mL of liquid LB media and plate a 50 μL aliquot on LB-agar plates containing the appropriate antibiotics (*see* **Note 5**).
f. Plate the remaining cell suspension onto LB-agar library plates (245 mm) containing the appropriate antibiotic selection marker and incubate the plates at 37°C overnight (*see* **Note 6**).

5. Carefully suspend the colonies from the agar plate into 2X 10 mL of LB medium. Collect the cell suspension and pellet the culture by centrifugation at 1500*g* at 4°C for 15 min.
6. Resuspend the pellet in a mix containing 1.5 mL of LB medium and 0.5 mL of glycerol.
7. Pipet 100 μL of the cells into an Eppendorf tube and pellet the suspension (*see* **Note 7**).
8. Remove the supernatant and isolate the plasmid DNA from the pellet, using Qiagen's QIAprep Mini Prep kit (for details see manufacturer's protocol).
9. Spectrophotometrically determine the final DNA concentration of all samples (*see* **Note 8**).
10. Set up 50 μL PCR reactions containing 1X *Taq* DNA polymerase buffer, 1.25 m*M* $MgCl_2$, 0.5–1 μ*M* of primer A and B (*see* **Note 9**), ~10 ng of linearized template vector, 5 units of *Taq* DNA polymerase, and 200 μM dNTP.
11. Amplify template over 30 cycles of PCR (94°C for 30 s, T_A for 30 s, 72°C for 1.2 min/kb) (*see* **Note 10**).
12. Check the quality of the PCR products by running 1–2 μL of reaction mixture on an agarose gel next to a suitable size marker. Purify the remaining reactions with Qiagen's QIAquick PCR purification kit.
13. Spectrophotometrically determine the final DNA concentration of all samples (*see* **Note 8**).

3.2. Reading Frame Selection of DNA Fragments

1. Digest 10 μg of the PCR product (DNA library) and 1.5 μg of pSALect vector with 20 units of *Nde*I and 20 units of *Spe*I restriction enzymes at 37°C for 6–8 h.
2. Dephosphorylate the pSALect vector by addition of 0.5 μL of shrimp alkaline phosphatase and continue incubation for another hour at 37°C.
3. Purify the reaction mixtures by agarose gel electrophoresis: Pour a 1% agarose gel with sufficiently large wells to load entire digests; Load digests in wells after mixing it with loading dye and run gel; Visualize the bands of DNA under UV light and excise product bands that correspond in size to the expected fragments with a razor blade; Recover DNA from the gel matrix, using Qiagen's QIAquick gel purification kit (for details see manufacturer's protocol) and elute DNA with 50 μL of Qiagen's EB buffer.
4. Ligate 10 μL of vector DNA with 10 μL of hybrid DNA in a 25-μL reaction with 1X T4 DNA ligase buffer and 6 Weiss units of T4 DNA ligase at room temperature overnight.
5. Transform aliquots of the ligation mixture into electrocompetent *E. coli* (*see* **Note 11**): Thaw one frozen 50-μL aliquot of competent cells for each sample by

storing the vials on ice; Add 4 µL of ligation mixture per aliquot of competent cells, then transfer the cell suspension into an ice-cold electroporation cuvet, and store the cuvet on ice for 1 min. Load the cuvet into the electroporator and apply an electrical pulse (for optimal settings of electroporator, contact manufacturer's manual); Remove cuvet from electroporator and immediately add 1 mL of SOC medium. Incubate the cell suspension at 37°C for 1 h. Dilute a two-microliter culture sample in 1 mL of liquid LB media and plate a 50 µL aliquot on LB-agar plates containing 50 µg/mL chloramphenicol (*see* **Note 5**); Plate the remaining cell suspension onto LB-agar library plates (245 × 245 mm) containing 50 µg/mL chloramphenicol and incubate the plates at 37°C overnight.

6. Carefully suspend the colonies from the agar plate into 2X 10 mL of LB medium and collect the cell suspension.
7. Measure the optical density (OD) of the cell suspension at 600 nm.
8. Dilute cell suspension to approx 1 × 10^6 cells/mL, using LB medium (*see* **Note 12**).
9. Plate 2 mL of cell culture on LB-agar library plates (245 × 245 mm) containing 100 µg/mL ampicillin and incubate the plates at room temperature for 2–4 d (*see* **Note 13**).
10. Dilute 2 µL of cell culture in 200 µL of LB medium, plate 50 µL aliquots on LB-agar plates containing either ampicillin (100 µg/mL) or chloramphenicol (50 µg/mL), and incubate the plates at room temperature for 2–4 d (*see* **Note 14**).
11. Carefully suspend the colonies from the agar plate into 2X 10 mL of LB medium. Collect the cell suspension and pellet the culture by centrifugation at 1500*g* at 4°C for 15 min.
12. Resuspend the pellet in a mix containing 1.5 mL of LB medium and 0.5 mL of glycerol.
13. Pipet 100 µL of the cells in an Eppendorf tube and pellet the suspension (*see* **Note 7**).
14. Remove supernatant and isolate the plasmid DNA from the pellet, using Qiagen's QIAprep Mini Prep kit (for details see manufacturer's protocol).
15. Spectrophotometrically determine the final DNA concentration of all samples (*see* **Note 8**).

At this point, the plasmid DNA can be directly used for DNA shuffling. (*See* related chapters in this volume.) Alternatively, the library of hybrid genes can be amplified with gene-specific primers and the PCR product used for the shuffling protocol (*see* **Note 15**).

4. Notes

1. In a typical experiment, we split the reaction mixture in two equal aliquots. One of the aliquots is mixed with 1 µL of size marker (100 bp-ladder) and glycerol is added to both samples. We prefer glycerol as a loading aid instead of regular loading dye, because of interference of DNA intensity and mobility at the positions of the dye colors in the gel. The two aliquots are then loaded next to one another on the gel and run at 50–80 V for 3–4 h (using BioRad's MiniSub Cell GT system). The endpoint of the electrophoresis is determined by the occasional

monitoring of DNA migration using UV light. An extended running time will result in better separation and will allow more narrow size selection.

2. The range of fragment sizes in the final product can be controlled by the length of time to run the electrophoresis and the size of the excised fragment. When selecting a narrow band width, we suggest to excise multiple bands above and below the target size. These fragments are then purified individually and their size distributions analyzed by subsequent ligation and DNA sequencing. It is our experience that size ranges of ±30 bp can be routinely achieved with a 4 kb vector, with a feasible limit of ±10 bp.
3. Crucial to the successful intramolecular ligation is to maintain an overall DNA concentration of <3 ng/µL *(7)*. Adjust the volume of the ligation reaction so that the DNA concentration falls below this threshold.
4. We have had consistently good results with strain DH5α-E. Transformation by electroporation is generally more efficient, generating larger libraries than chemical transformation, however either method works for transforming incremental truncation libraries.
5. These plates serve as controls for the library size. The number of colonies on the library plate(s) can be estimated by counting the control plates and multiplying the number of colonies by a factor of 10,000.
6. Library sizes can range from a few hundred colonies to several million members. We recommend to estimate the theoretical library size (theoretical size = $N*n$; maximum = N^2) whereby N= the number of nucleotides overlapping between the two parental genes or gene fragments, and n= the base pair range after size selection. For example, consider an incremental truncation library between two genes with a 150 bp overlapping region (N= 150) size-selected for fragments of parental size plus-minus 20 nucleotides (n= 40). The resulting theoretical library size would thus be 6000 members. For a maximal probability of representation of each crossover, the number of transformants should be approx 5–10 times the theoretical library size. With respect to the above example, the library plate should consist of at least 30,000 colonies.
7. We recommend to flash-freeze and store the remaining cell suspension in 100 to 200-µL aliquots at –80°C.
8. We use the following rule for concentration measurements: 1 A_{260} = 50 µg DNA.
9. Primer A and B are gene-specific forward- and reverse-primers. The PCR step is important to prepare the library for the subcloning into the pSALect vector. First, the primer design must mutate the 3'-terminal Stop-codon of the hybrid gene to a glycine codon (GGA). Simultaneously, the necessary *Nde*I site at the 5'-end of the sequence and the *Spe*I site on the 3'-end can be introduced with these primers.
10. The optimal annealing temperature for primer A and B must be determined prior to the amplification. We recommend primers of 20–25 bp length with T_A of 55–60°C.
11. We have had consistently good results with *E. coli* strain DH5α-E. Transformation by electroporation is generally more efficient, generating larger libraries than chemical transformation, however either method works.

12. We use the following rule as a guideline to calculate the cell density in cultures; 1 $OD_{600} = 1 \times 10^8$ cells/mL.
13. The rate of cell growth varies widely. We obtained the best libraries by incubation of the library plates at room temperature. The lower temperature accommodates the expression of slow-folding proteins.
14. The diluted plates serve as controls. Plating approximately 500 colony forming units (cfu) per plate, all of them should grow in the presence of chloramphenicol but only a fraction will appear on the ampicillin-containing plate. For quality control, we recommend selecting random colonies from the ampicillin plate, growing them in liquid medium overnight, and isolating the plasmid DNA for sequence analysis to confirm that the crossovers they contain are in-frame.
15. Remember that the Stop codon needs to be reintroduced. This can be done during the PCR step prior to DNA shuffling. Alternatively, it can be reintroduced after the primerless reassembly step when amplifying with outside primers.

Acknowledgments

The described protocols were developed and implemented as part of the authors' postdoctoral work in Stephen J. Benkovic's laboratory at the Pennsylvania State University. We would like to acknowledge and thank him for his invaluable support and guidance.

References

1. Ostermeier, M., Shim, J. H., and Benkovic, S. J. (1999) A combinatorial approach to hybrid enzymes independent of DNA homology. *Nat. Biotechnol.* **17,** 1205–1209.
2. Stemmer, W.P. (1994) Rapid evolution of a protein in vitro by DNA shuffling. *Nature* **370,** 389–391.
3. Lutz, S., Ostermeier, M., Moore, G.L., Maranas, C.D., and Benkovic, S. J. (2001) Creating multiple-crossover DNA libraries independent of sequence identity. *Proc. Natl. Acad. Aci. USA* **98,** 11,248–11,253.
4. Lutz, S., Fast, W., and Benkovic, S. J. (2003) A universal, vector-based system for nucleic acid reading frame selection. *Protein Eng.* in press.
5. Waldo, S., Standish, B. M., Berendzen, J., and Terwilliger, T.C. (1999) Rapid protein-folding assay using green fluorescent protein. *Nat. Biotechnol.* **17,** 456–460.
6. Sieber, V., Martinez, C. A., and Arnold, F. H. (2001) Libraries of hybrid proteins from distantly related sequences. *Nat. Biotechnol.* **19,** 456–460.
7. Sambrook, J., Fritsch, E. F., and Maniatis, T. (1989) *Molecular Cloning: A Laboratory Manual, 2nd ed.*, Cold Spring Harbor Laboratory, Cold Spring Harbor, NY.

18

Sequence Homology-Independent Protein Recombination (SHIPREC)

Andrew K. Udit, Jonathan J. Silberg, and Volker Sieber

1. Introduction

Genomic recombination is widely recognized as the principal mechanism by which proteins evolve new functions *(1)*. Having realized the importance of recombination in evolution, scientists have developed a variety of methods to mimic this phenomenon in the lab to create libraries of gene chimeras. While the ideal technique would allow for the recombination of any parental genes, most available methods are limited to recombining closely related sequences, i.e., those with ≥70% sequence identity, resulting in crossovers biased towards regions of high identity. This occurs because most recombination methods, e.g., Stemmer-shuffling *(2)*, StEP *(3)*, RACHITT *(4)*, and in vivo methods *(5)*, rely on template switching to generate chimeras. However, two methods have recently been described which are capable of generating chimeric libraries independent of sequence identity. Sequence homology-independent protein recombination (SHIPREC) *(6)* is described in this chapter; another method, incremental truncation for the creation of hybrid enzymes (ITCHY) *(7)*, is also described in this volume. Both methods are capable of generating chimeric libraries containing all possible single crossovers between the two parental genes.

SHIPREC results in a library of chimeras in which the hybrids generated retain proper sequence alignment with the parents. When parental genes are chosen that encode homologous proteins, this type of recombination can produce a library of chimeric proteins in which the crossovers occur at structurally related sites. The general SHIPREC methodology is shown in **Fig. 1**. The starting material is a fusion of the two genes of interest, with the C-terminus of the first gene and the *N*-terminus of the second gene joined through a small linker containing a unique restriction site (e.g., *Pst*I). The gene fusion is then ran-

From: *Methods in Molecular Biology, vol. 231: Directed Evolution Library Creation: Methods and Protocols*
Edited by: F. H. Arnold and G. Georgiou © Humana Press Inc., Totowa, NJ

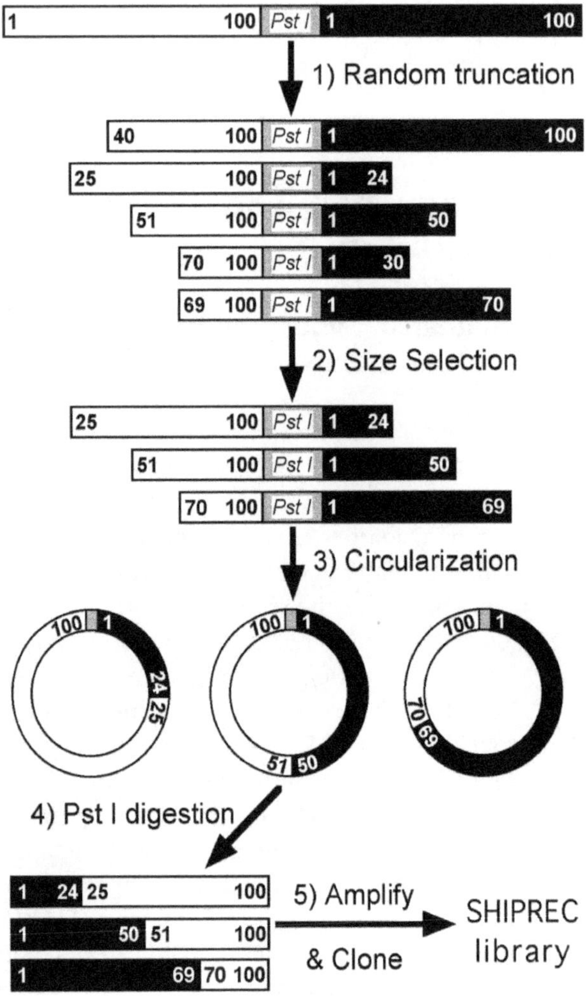

Fig. 1. SHIPREC overview. A gene fusion comprised of the two parental genes connected by a unique restriction site is constructed. (**1**) This fusion is randomly fragmented using DNase I and S1 nuclease to produce a library of gene fusions exhibiting varying length and containing blunt ends. (**2**) Gene fusions corresponding to the length of the parental genes are isolated using gel electrophoresis and separated from the random digest pool. (**3**) Single-gene length fragments are circularized by intramolecular blunt-end ligation. (**4**) Circular DNA is linearized by treatment with a restriction endonuclease that cuts in the linker that separates 3' and 5' ends of the original parental genes. This yields a library of chimeric genes that encode for proteins with a *N*-terminal region originating from Parent B and a C-terminal region originating from Parent A. (**5**) The chimeras are amplified and cloned into an appropriate vector for screening/selection.

domly truncated using DNase I and S1 nuclease, creating a library of fusions of varying size. From this library, DNA corresponding to the length of the parental genes is isolated and subsequently circularized. The size selection ensures that the circularization produces chimeras which retain proper sequence alignment with the parental genes. The resulting chimeras are linearized by cleaving the unique restriction site in the linker (i.e., *Pst*I), and the library is cloned into an appropriate vector for screening or selection. In order to generate all possible single-crossover chimeras, SHIPREC must be performed twice starting with both possible parental gene fusions, e.g., A–B and B–A.

2. Materials

1. Luria Bertani (LB) broth and appropriate antibiotic.
2. Oligonucleotide primers, 10 μ*M* in ddH_2O.
3. Ligation buffer, 5X stock: 250 m*M* Tris-HCl, pH 7.5, 50 m*M* $MgCl_2$, 10 m*M* dithiothreitol (DTT), 125 mg/mL bovine serum albumin (BSA).
4. 1.0 m*M* ATP.
5. 1% Ethidium bromide in water.
6. 100 m*M* $MnCl_2$.
7. 0.5 *M* ethylene diamine tetraacetic acid (EDTA) pH 8.0.
8. Bovine serum albumin (BSA), 1 mg/mL.
9. 100 m*M* Tris-HCl, pH 7.5.
10. Deoxyribonucleotide triphosphate mix (dNTP): 10 m*M* each dATP, dCTP, dGTP, dTTP.
11. Restriction endonucleases.
12. High activity T4 DNA ligase, 2000 U/μL (New England Biolabs, Beverly, MA).
13. S1 nuclease, 100 U/μL (Fermentas, Hanover, MD).
14. Deoxyribonuclease I, RNase-free (Sigma, St. Louis, MO).
15. *Vent* DNA Polymerase, 2 U/μL (New England Biolabs, Beverly, MA).
16. Parental genes.
17. High copy bacterial plasmid (e.g., pBluescript, pSTBlue 1, pBC).
18. *Escherichia coli* XL1-Blue competent cells (Stratagene, La Jolla, CA).
19. 37°C incubator and 55°C dry bath.
20. Agarose gels, electrophoresis equipment, and UV transilluminator.
21. DNA molecular weight standards.
22. Zymoclean Gel DNA Recovery kit (Zymo Research, Orange, CA).
23. DNA Clean and Concentrator-25 (Zymo Research, Orange, CA).
24. Spin-X Columns (Corning Costar Co., Corning, NY).
25. QIAGEN Plasmid Midi kit (Qiagen, Valencia, CA).
26. Scalpels.

3. Methods

The methods below outline 1) the fusion of two genes, 2) generation of a library of randomly truncated gene fusions, 3) isolation of parental length gene

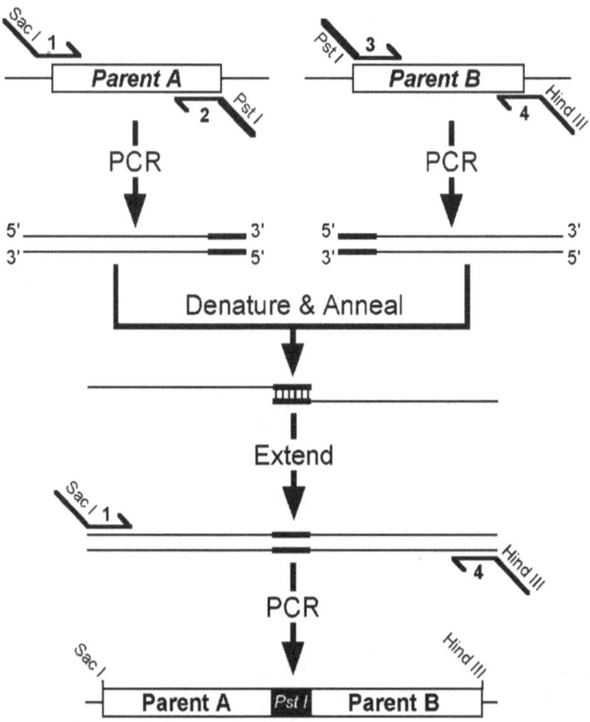

Fig. 2. Fusing parental genes using splicing by overlap extension. The two parental genes are individually PCR amplified using primers that contain complementary overhangs encoding a unique restriction site, *Pst*I. The two amplified products are mixed, denatured, and a second PCR reaction is performed using only flanking primers that contain restriction sites for cloning, *Sac*I and *Hin*dIII.

fusions, 4) circular permutation of the isolated genes, and 5) amplification and cloning of the resulting gene chimeras.

3.1. Gene Fusion Assembly

A gene fusion comprised of the two parental genes connected through an oligo linker is first constructed using splicing by overlap extension (SOEing) *(8)*. This procedure is outlined in **Fig. 2** and **Subheadings 3.1.1–3.1.4.** The basic principle involves PCR amplifying both parental genes with primers containing complementary overhangs that contain a unique restriction site (i.e., *Pst*I). The two amplified genes are then mixed, annealed, and a second PCR reaction is performed using only flanking primers (primers 1 and 4; *see* **Fig. 2**) that contain restriction sites for cloning (i.e., *Sac*I and *Hin*dIII). The resulting gene fusion is cloned into an expression vector, verified by sequencing, and amplified to generate sufficient DNA for subsequent steps.

3.1.1. Primer Design

Four primers are required to fuse two parental genes together in a single orientation (*see* **Fig. 2**). Primers 1 and 2 should be designed for amplifying parent A, and primers 3 and 4 should allow for amplification of parent B. Primers 2 and 3, in addition, should contain complementary overhangs that contain unique restriction sites not found in either gene (*see* **Note 1**). Furthermore, primers 1 and 4 should contain appropriate restriction sites to allow for cloning of the gene fusion into a high copy bacterial vector (*see* **Note 2**).

3.1.2. Primary PCR

1. Amplify each parental gene in separate PCR reactions. Each reaction should contain 10 µL 10X *Vent* Buffer, 2 µL dNTP mix, 5 µL of each primer, 10 ng template, and 1 µL *Vent*. Water should be added to bring the total volume to 100 µL.
2. Run an appropriate PCR cycle for the primers chosen (*see* **Note 3**).
3. Isolate the amplified genes using a 0.8% Tris-acetate EDTA (TAE) agarose gel containing 0.001% ethidium bromide using standard techniques *(9)*.
4. View the gel with an UV transilluminator under the preparative intensity and excise the bands exhibiting the correct molecular weight.
5. Purify the DNA from the excised agarose using the Zymoclean Gel DNA Recovery kit. Elute the DNA two times, using 20 µL ddH_2O per elution (*see* **Note 4**).
6. Determine the concentration of the purified DNA.

3.1.3. Secondary PCR

1. The secondary PCR reaction should contain 10 µL 10X *Vent* Buffer, 2 µL dNTP mix, 5 µL of primer 1, 5 µL of primer 4, 10 ng of parents A and B from the primary PCR, and 1 µL *Vent*. Water should be added to bring the total volume to 100 µL (*see* **Note 5**).
2. Run an appropriate PCR cycle for the primers chosen.
3. Gel purify the gene fusion as described in **Subheading 3.1.2.**

3.1.4. Cloning

Clone the gene fusion into a high copy bacterial vector using standard methods *(9)*. Transform *E. coli* with this vector, and verify the inserted gene fusion by sequencing. Grow up a 50 mL LB culture of *E. coli* harboring the plasmid, and purify the vector from this culture using the QIAGEN Plasmid Midi kit. This scale of amplification should yield ~100 µg of plasmid containing the insert which should be sufficient material for multiple SHIPREC experiments (*see* **Note 6**).

3.1.5. Purification of Gene Fusion

1. Digest all 100 µg of the plasmid DNA obtained from the Midi-kit with restriction enzymes that excise the gene fusion without cleaving it; use the restriction sites employed for cloning in **Subheading 3.1.4.**

2. Gel purify the gene fusion as described in **Subheading 3.1.2.** (*see* **Note 7**).
3. Determine the concentration of the purified gene fusion. This step should yield a DNA stock of ~50 μg.

3.2. Random Truncation of the Gene Fusions

A random library of truncated genes is generated by digesting the purified gene fusion with DNase I in the presence of Mn^{+2} ***(10)***. Under these conditions, DNase I cleaves dsDNA and produces an ensemble of random length fragments containing staggered ends (5' and 3' overhangs) as well as double stranded DNA with single stranded nicks. S1 nuclease is then used to cleave DNA at the nicks and convert staggered ends to blunt ends (*see* **Note 8**) ***(11)***.

3.2.1. DNase I Digestion

1. The digestion reaction consists of 5 μg purified gene fusion, 25 μL 100 m*M* Tris-HCl, pH 7.5, 10 μL 100 m*M* $MnCl_2$, 5 μL 1 mg/mL BSA, and 1 μL of DNase I (50 mU/μL stock; *see* **Note 9**). Water should be added to bring the total volume to 100 μL.
2. Incubate the reaction at room temperature.
3. Terminate aliquots of the reaction at 1, 2, 4, 8, and 16 min by taking 20 μL of the reaction and adding it to a different tube containing 2 μL of 0.5 *M* EDTA. Mix thoroughly and place the terminated aliquots on ice (*see* **Note 10**).
4. Purify the DNA from each time point separately using the DNA Clean and Concentrator-25 kit (*see* **Note 4**). The results of an ideal digestion are shown in **Fig. 3A**.

3.2.2. S1 Nuclease Treatment

1. For each DNase I time point, mix 35 μL purified DNA, 5.4 μL 7.4X S1 buffer, and 0.5 μL S1 Nuclease.
2. Incubate at room temperature (23°C) for 20 min (*see* **Note 11**).
3. Terminate each reaction by purifying the DNA with the DNA Clean and Concentrator-25 kit (*see* **Notes 4** and **12**). The results of a typical S1 treatment are shown in **Fig. 3B**.

3.3. Size Selection

Agarose gel electrophoresis is used to separate the gene library based on size. DNA corresponding to the approximate length of the individual parental genes is excised and purified.

1. Prepare a 0.8% TAE agarose gel containing 0.001% ethidium bromide and sufficient lanes for all DNA samples and molecular weight standards using standard techniques ***(9)***.
2. Load similar volumes and amounts of the digested DNA and molecular weight standards into each lane (*see* **Note 13**).
3. Run the gel until the molecular weight markers are sufficiently separated.

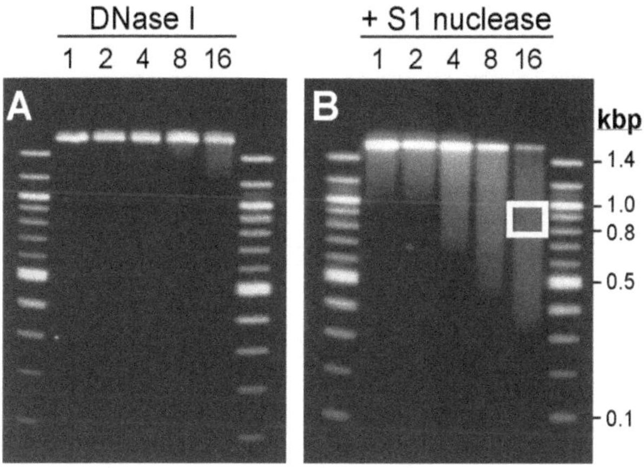

Fig. 3. Random digestion of a 1.8 kb gene fusion. **(A)** The results of 1, 2, 4, 8, and 16 min DNase I digestions. **(B)** S1 Nuclease treatment of DNase I digested DNA for each time point. Incubation with S1 Nuclease results in further digestion, a consequence of the enzyme cleaving at the sites where DNase I has nicked the template DNA. The box shown represents the molecular weight range to be excised during the size selection for parental genes that are 900 bp in length.

4. View the gel with an UV transilluminator using the preparative intensity. Each reaction should appear as a smear (*see* **Note 14**).
5. Using the molecular weight markers as a guide, excise the DNA from the smear that corresponds to a range of molecular weights centered around the size of the parental genes (*see* **Note 15** and **Fig. 3B**).
6. Purify the DNA from the excised agarose using the Zymoclean Gel DNA Recovery kit and determine the concentration of the purified sample.

3.4. Circularization

A library of circular permutated gene fusions is created by blunt ligation *(12)*, fusing the truncated *C*-terminus of the second gene with the truncated *N*-terminus of the first gene. The circularized DNA is concentrated and purified, and the library of circularized genes is linearized by digesting with the unique restriction site in the linker of the gene fusion (i.e., *Pst*I in **Fig. 1**).

1. Ligate the size selected DNA in a 100 μL reaction containing 200 ng purified DNA, 20 μL 5X Ligase buffer, 5 μL 1.0 m*M* ATP, 2 μL high activity ligase, and an appropriate volume of ddH_2O (*see* **Note 16**).
2. Incubate the reaction overnight at 16°C.
3. Terminate the ligation reaction by purifying the DNA with the DNA Clean and Concentrator-25 kit. Elute the DNA from the column provided with the kit twice with 17.5 μL ddH_2O per elution.

4. Linearize the circular permutated library by digesting the purified DNA with a restriction enzyme that cleaves in the original gene fusion linker (i.e., PstI in **Fig. 1**).

3.5. Amplification and Cloning

The library of chimeras is amplified by PCR, the DNA corresponding to the parental length genes is isolated, and this DNA is cloned into an expression vector.

3.5.1. PCR Amplification

Amplify the chimeras with the primers used to generate the original gene fusion. Perform this PCR as described in **Subheading 3.1.2..**, but use 4 μL of the digested DNA from **Subheading 3.4.** as the template and use the appropriate primers to amplify chimeras containing a single crossover (*see* **Note 17**). Note that if you started with the gene fusion *gene A–gene B*, then your chimera will be in the reverse orientation (i.e., *gene B-gene A*).

3.5.2. Purification of Chimeras

Amplification of a SHIPREC library can result in a smear centered around the desired products. Therefore, to ensure that the amplified library contains mostly chimeras of the desired range of sizes, a second size selection should be performed as described in **Subheading 3.3.** (*see* **Note 18**).

3.5.3. Cloning the Library

Clone the library into an appropriate vector, using the restriction sites that were placed in the primers. Restriction digestion of the library and plasmid, and subsequent cloning should follow standard procedures *(9)*. The pieces can then be ligated and transformed into *E. coli*, and the transformed cells can be grown up. The library can now be analyzed, screened, or selected (*see* **Notes 19** and **20**).

4. Notes

1. Primers 2 and 3 should contain complementary overhangs with a melting temperature of >60°C. We have successfully used the sequence ACTAGT GGATCCCTGCAGGAATTC for primer overhangs; this linker codes for four restriction sites (*Spe*I, *Bam*HI, *Pst*I, and *Eco*RI) to choose from when linearizing the circularized DNA in **Subheading 3.4.**
2. If both gene orientations are to be constructed and used (i.e., A–B and B–A), we recommend engineering identical cloning restriction sites into the corresponding flanking primers for both gene fusions. For example, primer 1 as shown in **Fig. 2** should always contain the same restriction site whether or not it is being used to generate fusion A–B or B–A; the same can be said for primer 4. Use of the same restriction sites in corresponding primers allows for the use of the same bacterial

vector for cloning the gene fusion, as well as the A–B and B–A library of chimeras. We recommend using *Sac*I for primer 1 and *Hin*dIII for primer 4 if your vector allows for this; we have had good experiences with these enzymes.

3. We find that 30 cycles using the following parameters give good yields with *Vent* when amplifying 1 kb genes: 94°C for 30 s (denaturation), 55°C for 1 min (annealing), and 72°C for 1 min (extension). However, the annealing temperature may have to be increased to prevent non-specific annealing, and the time for extension may need to be extended if amplifying larger templates.
4. To obtain the highest yields possible, we recommend using the mini-spin columns provided with the DNA Clean and Concentrator-25 kit from Zymo Research. In addition, we recommend eluting the DNA from the column twice using 20 μL of ddH_2O each time rather than a single elution with 35 μL, as recommended by the manufacturer.
5. Typically, varying template concentration does not significantly affect the yields from the secondary PCR. Similar results are obtained using 2–100 ng of template.
6. The PCR amplified gene fusion from **Subheading 3.1.3.** could be used to perform SHIPREC, but we recommend cloning the fusion into a vector and confirming the sequence prior to proceeding.
7. If you wish to purify larger quantities of DNA, we recommend a different protocol for extracting the DNA from the agarose, as the Zymo Research kit is not as efficient with larger quantities of DNA. Take the piece of excised agarose and slice it up into many small pieces using a scalpel. Place these pieces into a Spin-X column, and centrifuge for 3 minutes at 10,000 rpm. Remove the effluent and wash the column by adding 250 μL ddH_2O to the spin cup and centrifuge again for 3 min at 10,000 rpm. Pool the effluents and ethanol precipitate the DNA using standard procedures *(9)*.
8. The DNA overhangs created by DNase I digestion are often staggered and cannot be ligated in **Subheading 3.4.** S1 nuclease treatment is essential for successful ligation; this treatment serves the dual purpose of blunting staggered ends and cleaving at single stranded nicks.
9. We recommend using RNase-free deoxyribonuclease I from Sigma Chemicals (cat. no. D7291). The enzyme is typically supplied as a ~100 U/μL stock. A 0.05 U/μL working stock of DNase I should be made in 100 m*M* Tris-HCl, pH 7.5 and 40% glycerol. The working stock should be aliquoted into several tubes, snap frozen in liquid nitrogen, and stored at –80°C. The aliquots should be used once and then discarded; this ensures reproducibility in experiments.
10. Taking multiple time points allows for easy and rapid optimization of the truncation process. If the time points chosen yield excessive digestion as observed in **Subheading 3.3.**, then use a lower concentration of DNase I and shorter time points. Conversely, if the digestion is not sufficient, then incubate the sample with DNase I for longer times. It is much easier to control the extent of DNase I digestion by varying the incubation time rather than using a range of DNase I concentrations.
11. Treat the sample with S1 nuclease for no more than 20 min. Excessive treatment of DNA with S1 nuclease will result in additional dsDNA cleavage.

12. The purpose of this step is to remove the glycerol from the S1 digestion, which is necessary to insure proper migration of the DNA during the electrophoresis in **Subheading 3.3.** S1 nuclease reactions typically contain ~8% glycerol, which can cause the sample to migrate differently than the standards, resulting in improper size selection.
13. Loading different amounts and volumes of DNA in adjacent lanes can result in aberrant migration, resulting in excision of the incorrect band from the DNA smear in **Subheading 3.3.**
14. A uniform smear lacking distinct bands should be observed. If any discrete bands other than the uncut gene fusion are seen, then the DNase I and S1 nuclease digestions should be repeated. Also, take note of the smear pattern; the ideal pattern has the DNA smear centered around the molecular weight of the single gene.
15. It is not necessary or possible to precisely excise from the gel only the DNA corresponding to the exact molecular weight of the parental genes. In fact, it is recommended to excise the entire band in the smear corresponding to ca. ±100 base pairs of the parental gene lengths to insure that the fragments of interest, those recombined at structurally homologous sites, are excised. This becomes even more desirable when one considers that the two genes being used are not likely to be exactly the same length, and so there is some ambiguity as to what length is designated as the size of the single gene. This step also allows the researcher a great deal of flexibility. Depending on the particular application, it may be desirable to select those fragments with either insertions or deletions.
16. This is by far the most difficult step in the protocol. Low concentrations of DNA (ca. 2 ng/μL) and ATP (50 μ*M*) are required to favor blunt end intramolecular ligation ***(9)***. Ligations performed using high levels of DNA result in concatamer formation, and excessively low concentrations do not give sufficient yields of circularized DNA. Furthermore, a high concentration ligase is needed to obtain adequate yields of the circularized product. Use of low activity ligase (400 U/μL) can also significantly decrease the efficiency of intramolecular circularization.
17. The restriction digest is not necessarily required before the PCR amplification. However, we have found that linearizing the library before PCR significantly improves the yield and quality of the amplified product.
18. In order for the subsequent cloning reaction to proceed efficiently, it is recommended to purify the PCR product to remove the primers and any other nonspecific products that may have resulted. Experience has shown that cloning a library into a vector can prove rather difficult; thus, it is necessary that all components for the cloning reaction be of the highest quality.
19. A bias in the distribution of crossovers can be introduced while performing SHIPREC. Therefore, the distribution of crossovers in the naïve (i.e., unselected) library should be examined to confirm that all possible variants are being generated. Probe hybridization can be used as a high throughput screen for crossover distribution ***(13)***; a protocol for this method can be found in this volume.
20. Two thirds of the variants in the library contain a shift in the reading frame at the crossover site. This results in an extra burden in the functional analysis of the

library. If no efficient functional selection of the chimeras is available, then a general selection scheme (i.e., fusing the library to chloramphenicol acetyl transferase or with the head proteins of filamentous phages) can be used ***(14,15)***. These protocols are outlined in detail in the second volume of this set.

References

1. Cui, Y., Wong, W. H., Bornberg-Bauer, E., and Chan, H.S. (2002) Recombinatoric exploration of novel folded structures: a heteropolymer- based model of protein evolutionary landscapes. *Proc. Natl. Acad. Sci. USA* **99,** 809–814.
2. Stemmer, W. P. (1994) DNA shuffling by random fragmentation and reassembly: in vitro recombination for molecular evolution. *Proc. Natl. Acad. Sci. USA* **91,** 10,747–10,751.
3. Zhao, H., Giver, L., Shao, Z., Affholter, J.A., and Arnold, F.H. (1998) Molecular evolution by staggered extension process (StEP) in vitro recombination. *Nat. Biotechnol.* **16,** 258-261.
4. Coco, W. M., Levinson, W. E., Cryt, M. J., et al. (2001) DNA shuffling method for generating highly recombined genes and evolved enzymes. *Nat. Biotechnol.* **19,** 354–359.
5. Volkov, A. A., Shao, Z., and Arnold, F. H. (1999) Recombination and chimeragenesis by in vitro heteroduplex formation and in vivo repair. *Nucleic Acids Res.* **27,** e18.
6. Sieber, V., Martinez, C.A. & Arnold, F.H. (2001) Libraries of hybrid proteins from distantly related sequences. *Nat. Biotechnol.* **19,** 456–460.
7. Ostermeier, M., Nixon, A. E., and Benkovic, S. J. (1999) Incremental truncation as a strategy in the engineering of novel biocatalysts. *Bioorg. Med. Chem.* **7,** 2139–2144.
8. Horton, R. M., Hunt, H. D., Ho, S. N., Pullen, J. K., and Pease, L. R. (1989) Engineering hybrid genes without the use of restriction enzymes: gene splicing by overlap extension. *Gene* **77,** 61–68.
9. Sambrook, J., Fritsch, E. F., and Maniatis, T. (1989) *Molecular Cloning, A Laboratory Manual*, Cold Spring Harbor Laboratory Press, Cold Spring Harbor, NY.
10. Melgar, E. and Goldthwait, D. A. (1968) Deoxyribonucleic acid nucleases. II The effects of metals on the mechanism of action of deoxyribonuclease I. *J. Biol. Chem.* **243,** 4409–4416.
11. Kroeker, W. D. and Kowalski, D. (1978) Gene-sized pieces produced by digestion of linear duplex DNA with mung bean nuclease. *Biochemistry* **17,** 3236–3243.
12. Graf, R. and Schachman, H.K. (1996) Random circular permutation of genes and expressed polypeptide chains: application of the method to the catalytic chains of aspartate transcarbamoylase. *Proc. Natl. Acad. Sci. USA* **93,** 11,591–11,596.
13. Joern, J. M., Meinhold, P., and Arnold, F. H. (2002) Analysis of shuffled gene libraries. *J. Mol. Biol.* **316,** 643–656.
14. Maxwell, K. L., Mittermaier, A. K., Forman-Kay, J. D., and Davidson, A. R. (1999) A simple in vivo assay for increased protein solubility. *Protein Sci.* **8,** 1908–1911.
15. Sieber, V., Pluckthun, A., and Schmid, F. X. (1998) Selecting proteins with improved stability by a phage-based method. *Nat. Biotechnol.* **16,** 955–960.

19

Producing Chimeric Genes by CLERY

In Vitro and In Vivo Recombination

Valérie Abécassis, Denis Pompon, and Gilles Truan

1. Introduction

Directed evolution usually starts from the analysis of variant genes. Among methods used to produce such diversity, recombination between members of a gene family is a frequent approach *(1)*. Building a library of mosaic genes can been achieved by fragmentation of parental genes and subsequent recombination by PCR without primers *(2,3)*. Cloning of the newly generated fragments need a ligation step followed by transformation of the universal cloning host: *Escherichia coli*. Here we describe a method combining a PCR-dependent reassembly of fragmented full expression vectors using optimized temperature cycles and an in vivo recombination and self-cloning in yeast. Cloning performed in yeast avoid the usual bias that could be introduced by ligation and propagation in *E. coli*, particularly any toxicity or counter-selection that would selectively apply to clones in the library. The method is illustrated by the construction of a combinatorial library between the human *CYP1A1* and the *CYP1A2* cDNA, which share 74% nucleotide sequence identity. Formation of at least 86% of mosaic genes was observed.

2. Materials

2.1. DNase I Digestion

1. Expression vectors p1A1/V60 *(4)* and p1A2/V60 *(5)* or equivalent.
2. DNase I, molecular biology grade (Grade II, Sigma-Aldrich).
3. Standard buffer for DNase I treatment: 50 m*M* Tris-HCl, pH 8.0, 10 m*M* $MnCl_2$. This buffer has to be prepared immediately before use to avoid manganese ion

From: *Methods in Molecular Biology, vol. 231: Directed Evolution Library Creation: Methods and Protocols*
Edited by: F. H. Arnold and G. Georgiou © Humana Press Inc., Totowa, NJ

oxidation and precipitation. A solution containing 100 m*M* $MnCl_2$ in water can be stored for months at 4°C.

4. PCR apparatus.
5. DNA ladder: pUC19 digested with BstEII or any suitable ladder as long as it provides small fragment sizing (under 100 bp).
6. Agarose and agarose gel electrophoresis equipment.
7. Centrisep column (Princeton Separation Inc., Adelphia, NJ) for DNase fragment purification.

2.2. Reassembly of the Fragmented DNA and Amplification of the Reassembled Genes

1. Forward and reverse primers for pYeDP60 expression cassette *(6)*: a 5'-primer located in the GAL10-CYC1 promoter region (5'-CGTGTATATAGCGT GGATGGCCAG-3') and a 3'-primer located in the PGK terminator region (5'-GCACCACCACCAGTAG-3'). Any alternate couple of forward and reverse primers can be used, as long as they amplify cDNA with common flanking regions of at least 50 bp (optimally 150–200 bp) for recombination purposes.
2. DA-free column (Millipore, Bedford, MA) for PCR product gel extraction.

2.3. Yeast Transformation

1. cDNA void pYeDP60 expression vector system (*see* **Fig. 1**) or any yeast expression vector containing suitable flanking regions *(4)*.
2. Yeast strain W303-1B (Mat a; ade2-1; his3, leu2, ura3, trp1, canR, cyr+) or any yeast strain with good transformability and recombination properties.
3. 10X TE buffer: 100 m*M* Tris-HCl, pH 7.5, 10 m*M* EDTA.
4. 10X lithium acetate buffer: 1 *M* lithium acetate.
5. 1X TE-lithium acetate buffer: 10 m*M* Tris-HCl, pH 7.5, 1 m*M* EDTA, 100 m*M* lithium acetate.
6. PEG 4000 solution in water (500 g/L in water). This solution should be filtered, aliquoted by 800 μL and kept at –20°C.
7. 50 μg of sonicated and heat-denatured salmon sperm DNA. Denature and keep on ice before use.
8. SWA6 medium *(7)* or any suitable selective yeast culture medium.
9. YPGA medium per liter: 10 g yeast extract; 10 g bactopeptone; 20 mg adenine and 20 g glucose or any rich yeast medium.

3. Methods

The methods described below outline 1) DNase I digestion, 2) reassembly of the fragmented DNA, 3) amplification of the reassembled genes, and 4) transformation of *S. cerevisiae*.

3.1. DNase I Digestion

1. Random fragmentation with DNase I in the presence of Mn^{2+} was realized with the modifications described by Lorimer *(8)* and Zhao *(3)*.

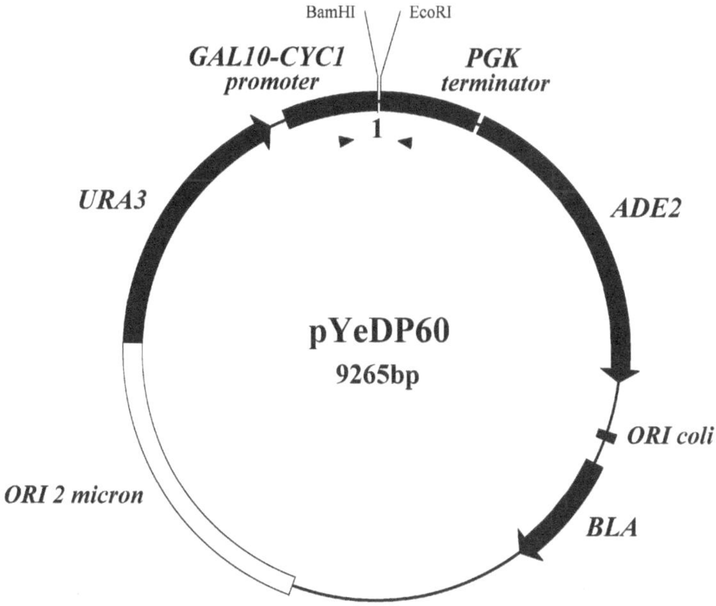

Fig. 1. pYeDP60 map. Arrows indicate the positions of the forward and reverse primers. *ADE2* and *URA3* represent the yeast genes used for transformation selection, *BLA* represents the *AmpR* gene for *E. coli* selection. *GAL10-CYC1 promoter* and *PGK terminator* represent the promoter and terminator used for cDNA expression of *CYP* ORFs. *ORI 2 micron* is the replication origin of the yeast 2μ plasmid, *ORI coli* represent the replication origin of pUC 19.

2. Dilute separately 7.5 μg of each plasmidic DNA (p1A1/V60 and p1A2/V60) in a buffer containing 50 m*M* Tris-HCl, pH 7.4, 10 m*M* $MnCl_2$ to a final volume of 120 μL and divide it into three tubes (40 μL each).
3. Add 40 μL of DNase I solution to each tube (DNase I is diluted in 50 m*M* Tris-HCl, pH 7.4, 10 m*M* $MnCl_2$). Usual quantities of DNase I are: 0.0112 U, 0.0056 U, and 0.0028 U of DNase I per digestion, but these values need to be adjusted (*see* **Note 1**).
4. Perform the digestion at 20°C for 10 min.
5. Terminate the reaction by heating at 90°C for 10 min.
6. Run 10 μL on a 2% agarose gel to verify the fragmentation pattern (*see* **Fig. 2**).
7. Purify the remaining volume (70 μL) of the DNase I reaction that gave the suitable fragmentation (*see* **Note 2**) using the Centrisep column equilibrated with Tris-HCl, 10 m*M*, pH 7.4 (without EDTA to avoid inhibition of the following PCR) and following the manufacturer's protocols. The elution volume should be around 70 μL.
8. These steps should always be performed without any time delay (except for agarose separation: leave the digestion on ice). If a solution containing $MnCl_2$ is

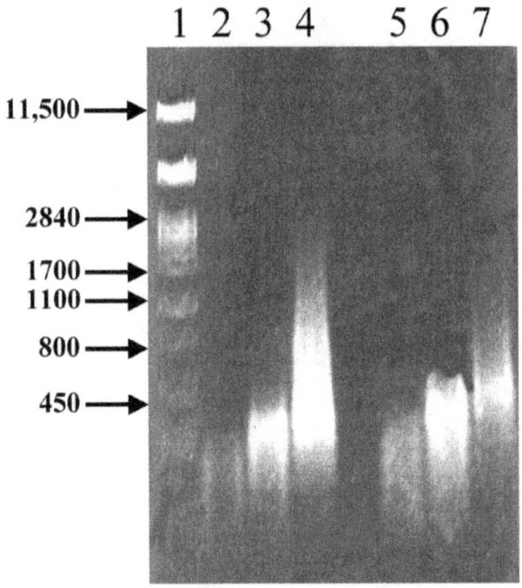

Fig. 2. Lane 1, DNA ladder; lanes 2, 3, 4 and 5, 6, 7 correspond to DNase I treated p1A1/V60 and p1A2/V60 respectively with decreasing concentration of DNase I. DNA fragments corresponding to the second DNase I concentration (lanes 3 and 6) were present at the expected size (200–300 bp) and were used for the reassembly.

frozen to –20°C, Mn ions will oxidize, precipitate (brownish color) and the material will then no longer be usable.

3.2. Reassembly Reaction

The procedure used was derived from the method described by Stemmer *(2,9)*.

1. Use 20 µL of the solution containing eluted fragments as templates (mix 10 µL of each digestion) in a 40 µL PCR reaction using 2.5 U of *Taq* polymerase, 4 µL of the 10X supplied buffer and 4 µL of 20 m*M* dNTPs. If more than two genes are fragmented, the maximum volume to be used in the PCR reaction is 20 µL (each fragmented DNA volume to be used will be 20/n, where n is the number of genes).
2. Perform the PCR program depicted in **Fig. 3A**.
3. Verify PCR reassembly with 5 µL of the reaction on a 1% agarose gel (*see* **Fig. 4**).

3.3. Amplification of the Reassembled Genes

1. Amplify 1 µL of the newly reassembled DNA in a 100 µL PCR reaction using 1 U of *Taq* polymerase in the supplied buffer and the forward and reverse pYeDP60 primers.
2. Perform the PCR program depicted in **Fig. 3B**.
3. Separate amplified fragments of the suitable size on a 1% agarose gel electrophoresis (*see* **Fig. 5**) (*see* **Note 3**).

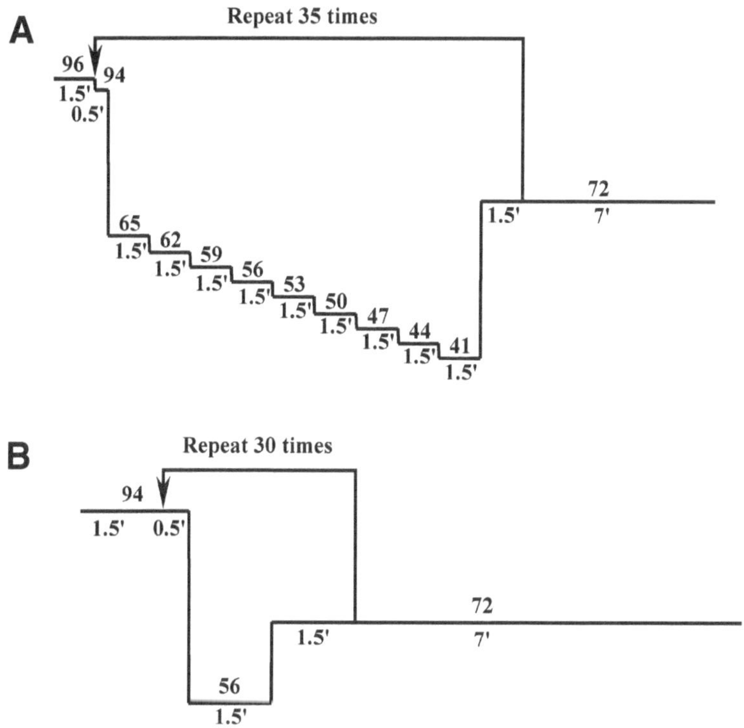

Fig. 3. Schemes of the different PCR programs. Times are in min (above lines), temperatures in °C (under lines). (**A**) reassembly PCR, (**B**) reamplification PCR.

4. Purify the amplified reassembled genes using DA-free column.
5. Quantify the purified DNAs using agarose gel electrophoresis or any other method.

3.4. Transformation of S. cerevisiae

The strategy takes advantage of a unique property of yeast: the capability to perform homeologous and homologous recombinations. Shuffled DNAs are inserted in the yeast expression vector pYeDP60 by gap-repair ***(10)***. This step constitutes a potential (depending on selected fragments) second round of DNA shuffling involving different molecular mechanisms ***(6,11)*** and allows direct cloning, expression and functional selection in yeast without the need for an intermediate cloning step in *E. coli*.

1. Linearize and purify pYeDP60 expression vector (*see* **Note 4**).
2. Perform an overnight preculture of W303-1B yeast strain in 5 mL YPGA at 28°C.
3. Dilute cells in 50 mL of YPGA medium to a final density of 2.10^6 cells/mL.

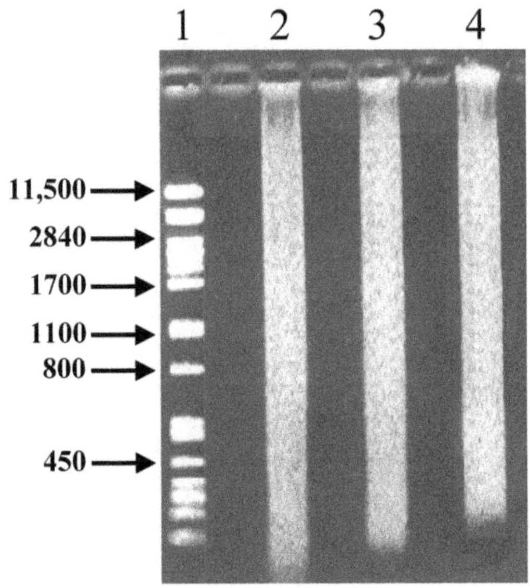

Fig. 4. Reassembly reaction migration performed on a 1% agarose gel. Lane 1, DNA ladder; lanes 2, 3 and 4 correspond to reassembly reactions between fragmented p1A1/V60 and p1A2/V60 for 0.0112, 0.028 and 0.056 U of DNAse I, respectively. Smears indicate reassembly processes, the lower limit corresponding to the lower limit from the DNase I digestion.

4. After six hours growth, wash the cells twice with sterile water and twice with a 1X TE-lithium acetate buffer.
5. Add 100 µL of 1X TE-lithium acetate buffer to the cells, mix and incubate 15 min at 28°C.
6. Add 50 µg of sonicated and heat-denatured salmon sperm DNA, 200–400 ng (or more if available) of reassembled-amplified DNA (insert), 0.025 µg (or more) of previously linearized pYeDP60 to 50 µL of cell suspension. Those quantities can obviously be larger to increase library size, keeping in mind that a 1:10 molar ratio between vector and insert is optimal. The total DNA volume should never exceed 5 µL, but scaling-up can be performed if needed keeping DNA solution volume smaller than 10% of the cell suspension volume.
7. Add 350 µL of freshly prepared PEG solution (800 µL PEG 40%, 100 µL 10X TE, 100 µL 10X lithium acetate).
8. Incubate the cell suspension with moderate shaking at 30°C for 30 min.
9. Incubate the cell suspension at 42°C for 20 min.
10. Pellet the cells (5 min at 5000*g*) and wash them with a 0.4% NaCl solution.
11. Pellet again the cells and add 200 µL of 0.4% NaCl solution, plate onto selective SWA6 medium and incubate at 28°C for 3–4 d.

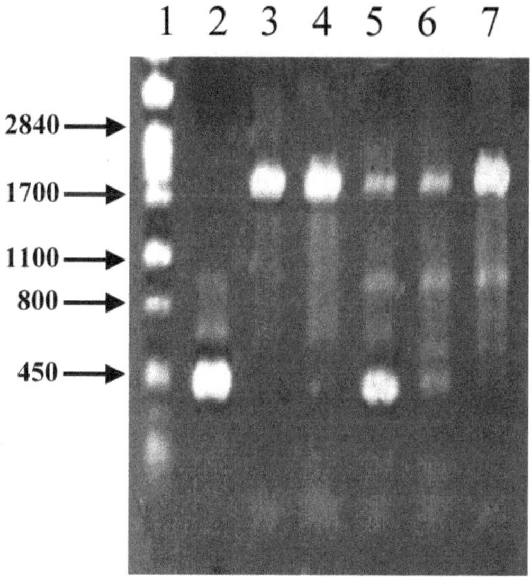

Fig. 5. Amplification reaction. Lane 1, DNA ladder; lanes 2, 3 and 4 correspond to the amplification with empty pYeDP60, p1A1/V60 and p1A2/V60, respectively; lanes 5, 6 and 7 correspond to the amplification with previously reassembled DNA as a matrix (**Fig. 4**, lanes 2, 3, 4 respectively). The three reassembled DNA gave the right PCR products size (*see* **Note 3**).

4. Notes

1. Because DNA preparations are generally different in term of quality, the rate of DNA fragmentation can be different from one sample to another. It is always useful to test different DNase I concentrations prior the real experiment. We found it suitable to make serial dilutions (by a factor of 2) of DNase I, starting from around 0.2 U down to 0.01 U per digestion instead of using different times of incubation.
2. Fragments around 200 bp to 300 bp (mean size) generally give the best results in reassembly. The use of small fragments (<150 bp, mean size) is not recommended, because they contain small oligonucleotides (<30 bp) that can inhibit the reassembly process. The use of large fragments (>500 bp) is not recommended either because of potential contamination by parental genes.
3. The amplification pattern following reassembly is indicative of the reassociation process. In **Fig. 5**, lane 7, the level of amplification probably denotes the presence of gene amplification from non-fragmented parental cDNAs. The presence of a contaminating band in lane 5 is also suspicious. The intermediate situation (lane 6) is therefore the best compromise.
4. The vector preparation is a critical step. With pYeDP60, we generally perform a double digestion with non-compatible sites to avoid recircularization in yeast.

Alternatively, if a single digestion is used, Klenow filling of the half sites will prevent recircularization. A good separation and elution of the linearized plasmid is also critical, because contamination by circular vector will generate a high proportion of empty vectors in the library. As a control, it is always advisable to transform yeast with the linearized vector preparation in the same conditions used for the library (using also 50 μL of cells). This will quantify the number of clones containing empty vectors in the library. Note that pYeDP60 can be substituted by any vector that that can be propagated in yeast and provides flanking regions compatible (identical) with the cDNA flanking regions. Particularly, the expression cassette can be adapted for functional expression into others hosts like mammalian cell cultures. In such case, following recombination in yeast, plasmids have to be extracted and the target host transformed directly with the extract or after reamplification in *E. coli*.

5. One needs to pay attention that yeast could be multi-transformed following recombination. Up to 5–10 different plasmids can sometime coexist in a rather stable fashion. This phenomenon may not affect selection and even be an advantage if the screening searches for a new function (i.e., not present in the parental proteins). If the screening is sensitive to interfering activities, subcloning is preferred. This can be achieved by pool extraction of plasmid from yeast, and transformation in *E. coli* or other hosts for segregation. Direct segregation in yeast is frequently inefficient except when selection pressure for desired function is available. Using a centromeric (low copy number) vector or performing the recombination step with a limiting DNA concentration can also be helpful if multiple transformation is a problem.

Acknowledgments

The work was supported in part by a grant from the GIP Hoechst Marion Roussel. Valérie Abécassis is a predoctoral fellow supported by a fellowship from the Ministère de la Recherche et de l'Enseignement Supérieur.

References

1. Crameri, A., Raillard, S. A., Bermudez, E., and Stemmer, W. P. (1998) DNA shuffling of a family of genes from diverse species accelerates directed evolution. *Nature* **391,** 288–291.
2. Stemmer, W. P. (1994) DNA shuffling by random fragmentation and reassembly: in vitro recombination for molecular evolution. *Proc. Natl. Acad. Sci. USA* **91,** 10,747–10,751.
3. Zhao, H. and Arnold, F. H. (1997) Optimization of DNA shuffling for high fidelity recombination. *Nucleic Acids Res.* **25,** 1307–1308.
4. Urban, P., Cullin, C., and Pompon, D. (1990) Maximizing the expression of mammalian cytochrome P-450 monooxygenase activities in yeast cells. *Biochimie* **72,** 463–472.
5. Bellamine, A., Gautier, J. C., Urban, P., and Pompon, D. (1994) Chimeras of the human cytochrome P450 1A family produced in yeast. Accumulation in micro-

somal membranes, enzyme kinetics and stability. *Eur. J. Biochem.* **225,** 1005–1013.

6. Abecassis, V., Pompon, D., and Truan, G. (2000) High efficiency family shuffling based on multi-step PCR and in vivo DNA recombination in yeast: statistical and functional analysis of a combinatorial library between human cytochrome P450 1A1 and 1A2. *Nucleic Acids Res.* **28,** E88.
7. Cullin, C. and Pompon, D. (1988) Synthesis of functional mouse cytochromes P-450 P1 and chimeric P-450 P3-1 in the yeast *Saccharomyces cerevisiae*. *Gene* **65,** 203–217.
8. Lorimer, I. A. and Pastan, I. (1995) Random recombination of antibody single chain Fv sequences after fragmentation with DNaseI in the presence of Mn2+. *Nucleic Acids Res.* **23,** 3067–3068.
9. Stemmer, W. P. (1994) Rapid evolution of a protein in vitro by DNA shuffling. *Nature* **370,** 389–391.
10. Pompon, D. and Nicolas, A. (1989) Protein engineering by cDNA recombination in yeasts: shuffling of mammalian cytochrome P-450 functions. *Gene* **83**, 15–24.
11. Mezard, C., Pompon, D., and Nicolas, A. (1992) Recombination between similar but not identical DNA sequences during yeast transformation occurs within short stretches of identity. *Cell* **70,** 659–670.

II

Analysis of Library Diversity

20

Analysis of Shuffled Libraries by Oligonucleotide Probe Hybridization

Peter Meinhold, John M. Joern, and Jonathan J. Silberg

1. Introduction

In vitro recombination is often used to generate genetic diversity for directed evolution. Recombination techniques that rely on fragment hybridization yield libraries with preferred crossover positions and, in some cases, a bias toward incorporation of one parent over the others. This limits the diversity, and thus the utility of the library. These biases vary depending on the technique used for recombination, the distribution of sequence similarities within the parental genes, and the efficiency by which the different parental genes are PCR amplified. To assess the diversity generated in a library of chimeras, sequences of a large number of chimeras are required. While DNA sequencing can yield this information, sequencing these genes is prohibitively expensive, especially when the genes being recombined are large. Oligonucleotide probe hybridization, in contrast, offers a cost-effective approach for obtaining information about library biases that allows for optimization of shuffling procedures. When coupled with functional information, this technique can provide information about the relationship between sequence and function *(1,2)*.

Probe hybridization in macroarray format, e.g., arrays of 96 or 384 chimeras, can determine the average number and relative locations of crossovers in chimeras generated from DNA shuffling. This method is capable of characterizing the structure of chimeric genes generated from parental genes exhibiting ≤90% sequence identity. The general methodology for probe hybridization analysis requires five main steps. First, a set of oligonucleotide probes that anneal to identical regions of the parental genes is generated, and the 3' end of the oligonucleotide probes are labeled with a tagged nucleotide using terminal transferase, e.g., dUTP-fluorescein. Second, bacteria are transformed with the

From: *Methods in Molecular Biology, vol. 231: Directed Evolution Library Creation: Methods and Protocols*
Edited by: F. H. Arnold and G. Georgiou © Humana Press Inc., Totowa, NJ

library of interest, arrayed on a nylon membrane, and allowed to grow. Once the colonies reach an acceptable size, the cells are lysed on the membrane, and the DNA remains bound to the nylon membrane where the colony grew. Third, each probe is hybridized to a membrane containing the arrayed gene library. Fourth, the variants containing hybridized probes are identified using a chemiluminescence system specific for the tagged nucleotide. Finally, the hybridization data are compiled in a simple spreadsheet to determine biases in the location of crossovers, and the average number of crossovers is determined using a statistical model ***(1)***.

While the described method contains numerous steps, good results are typically easily achieved without significant optimization.

2. Materials

2.1. Probe Design and Labeling

1. Oligonucleotide probes.
2. Terminal transferase, 8 U/μL and buffer.
3. Fluorescein-11-dUTP (*see* **Note 1**).
4. 37°C incubator.

2.2. Colony Blotting

1. *Escherichia coli* competent cells.
2. Gene library.
3. Luria Bertani (LB) broth and appropriate antibiotic.
4. Sterile 384-well plate with lid.
5. 384-pin replicator for the transfer of approx 100 nL.
6. Hybond-N+ nylon membrane, roll of 0.3 × 3 m (Amersham-Pharmacia Biotech; Piscataway, NJ).
7. Agar-plates containing appropriate antibiotic, 22 × 22 cm.
8. Tweezers with flat head.
9. Whatmann 3MM paper.
10. 96-well plate lids.
11. Denaturing solution: 1.5 *M* NaCl, 0.5 *M* NaOH.
12. Neutralizing solution: 1.5 *M* NaCl, 0.5 *M* Tris-HCl, pH 7.2, 10 m*M* EDTA.
13. SSC buffer, 20X stock: 3 *M* NaCl, 0.3 *M* sodium citrate, pH 7.0.
14. UV Crosslinker.

2.3. Hybridization

1. Hybridization oven.
2. Rollerbottles for oven, 35 × 150 mm.
3. Labeled probes.
4. Sodium dodecyl sulfate (SDS), 5% (w/v) in water.
5. Liquid blocking reagent (Amersham Pharmacia Biotech).

6. Hybridization buffer: 5X SSC, 0.1% SDS, 0.5% dextran sulfate (avg. MW= 500,000), 20-fold dilution of Liquid Block at –20°C.
7. Wash buffer: 5X SSC and 0.1% SDS, mix immediately before use.
8. Stringency buffer: 1X SSC and 0.1% SDS, mix immediately before use.

2.4. Detection

1. Buffer A: 100 m*M* Tris-HCl, pH 9.5, 300 m*M* NaCl (*see* **Note 2**).
2. Bovine serum albumin (BSA) fraction V.
3. Polyoxyethylene sorbitan monolaurate (Tween 20).
4. Orbital shaker.
5. Anti-fluorescein antibody conjugated to alkaline phosphatase (*see* **Note 3**).
6. CDP-Star detection reagent.
7. Saran wrap.
8. Biorad FluorS MultiImager (*see* **Note 4**).

3. Methods

The methods provided below describe the 1) design and labeling of gene-specific oligonucleotide probes, 2) arraying of chimeric genes on a membrane, 3) hybridization of probes to arrayed DNA, 4) detection of hybridized probes, and 5) analysis of hybridization data.

3.1. Probe Generation

A set of oligonucleotide probes should be designed to bind to each parental gene at multiple sites evenly distributed throughout the gene. In addition, probes for each parent should be located in identical regions of the respective genes.

3.1.1. Probe Design

When designing probes the following issues should be taken into consideration.

1. Initially at least four probe sites should be used. If the number of crossovers is expected to be higher than about five, more sites may be required to adequately assess library characteristics (*see* **Note 5**).
2. Probes should bind to regions of low sequence identity among the parental genes. It is also important that they bind to identical positions in the parental genes (*see* **Note 6**).
3. All probes should exhibit a calculated melting temperature (T_m) of ~62°C (*see* **Note 7**). If all probes exhibit similar annealing strength, then similar conditions can be used for hybridizing each probe.
4. The vector containing the library should be examined to ensure that probes do not bind to regions outside the gene of interest (*see* **Note 8**).
5. Inverted repeats should be avoided.
6. The G/C content should be 40–60%, and the termini should end in a G or C.

3.1.2. Probe Labeling

Label the 3' end of the probes with a tag that can be easily identified after hybridization. This can be rapidly performed using a modified nucleotide, e.g., fluorescein-11-dUTP, and Terminal transferase using standard procedures *(3)*.

1. Label each probe separately in a 80-μL reaction containing 100 pmoles probe, 32 U Terminal transferase, 60 μ*M* fluorescein-11-dUTP, and an appropriate volume of cacodylate buffer provided with the enzyme.
2. Incubate the reactions at 37°C for 90 min.
3. Store labeled probes at –20°C in the dark (*see* **Note 9**).

3.2. Colony Blotting

The library can be rapidly arrayed on a membrane by growing colonies transformed with the library in a 384-well plate (*see* **Note 10**), replicating the plate onto a nylon membrane placed on an agar plate (*see* **Note 11**), and growing the colonies directly on the membrane. The DNA can be denatured and immobilized on the membrane in an arrayed format by lysing the cells on the membrane, washing away the cellular debris, and crosslinking the DNA by UV irradiation.

3.2.1. Arraying Colonies

1. Pick freshly transformed colonies into a 384-well-plate containing 70 uL of media per well containing the appropriate antibiotic (*see* **Note 12**). As positive controls, bacteria transformed with the respective parental genes should be included in the plate, two wells per parent, at the edges of the blot. As negative controls, two wells should contain medium only.
2. Place the lid on the plate, wrap it in Saran wrap, and incubate overnight at 37°C, shaking at 270 rpm (*see* **Note 13**).
3. Cut the Hybond-N+ into sheets that are 7.5 × 11.5 cm and place on a LB- or M9 agar plate containing 2% (w/v) agar (*see* **Note 14**). Four membranes fit on 22 × 22-cm agar plates.
4. Replicate the cultures arrayed in the 384-well plate onto the membrane using the 384-pin replicator (*see* **Note 15**) and label the orientation of the array on the membrane, i.e. mark next to colony A1 with a pen.
5. Incubate at 37°C until colonies are 1–2 mm in diameter (*see* **Note 16**).

3.2.2. Lysing and Denaturing

1. Remove the membrane from the agar plate using tweezers and place, colony side up, on a pad of absorbent filter paper soaked in denaturing solution (*see* **Note 17**). The lids from 96-well plates are a convenient container for holding the filter paper and nylon membrane.
2. Incubate for 7 min at room temperature.
3. Transfer the membrane, colony-side up, to a pad of 3MM absorbent filter paper soaked in neutralizing solution (*see* **Note 17**).

4. Incubate for 3 min at room temperature.
5. Repeat the neutralization step with a fresh pad of filter paper.
6. Wash the membrane in 50 mL of 2X SSC. Both sides of the membrane should be wetted completely.
7. Using clean lab gloves, wipe the colony debris off the membrane. This drastically reduces any background signal. It does not decrease the signal intensity (*see* **Note 18**).
8. Wash the membrane in 50 mL 2X SSC.
9. Transfer the membrane, colony-side up, to an absorbent paper towel and dry at room temperature for 15 min.

3.2.3. Crosslinking

Crosslink the arrayed DNA to the nylon membrane using standard procedures, i.e., by UV-crosslinking or baking in a vacuum oven at 80°C *(3)*.

3.3. Hybridization

3.3.1. Optimization

Prior to hybridizing probes to an arrayed library, the conditions for hybridization should be optimized using the parental genes as controls. This requires performing hybridization experiments using a small array of the parental genes under different conditions, i.e., over a range of hybridization temperatures and stringency of wash salt concentrations (*see* **Notes 19** and **20**). Under optimal hybridization conditions, a probe will hybridize only to the parental gene that it was designed to bind and will not cross hybridize with any of the other genes.

3.3.2. Hybridization

1. Place the blots DNA-side up into 150 mL roller bottles and add 18 mL of hybridization buffer.
2. Incubate the blots for 1 h at 8 rpm in the hybridization oven at a temperature, which is 5°C lower than the calculated T_m of the probes.
3. Add 90 ng (~ 11 µL) of the labeled probe so that the final concentration of probe is 5 ng/mL.
4. Incubate the blots with probe in the hybridization oven for 2–12 h at 8 rpm at a temperature corresponding to T_m –5°C (*see* **Note 19**).
5. Remove the blots from the hybridization buffer and place in a clean roller bottle containing 120 mL wash buffer.
6. Incubate at room temperature for 5 min with gentle agitation.
7. Repeat with fresh buffer.
8. Discard wash buffer and cover the blots with an excess volume of stringency buffer.
9. Incubate in the hybridization oven for 15 min at 8 rpm using a temperature that decreases the signal of non-specific binding without significantly affecting the

desired signal (*see* **Note 20**). Stringency increases with temperature and decreases with salt concentration.

10. Repeat with fresh buffer if necessary.
11. The blots can be stored at 4°C overnight if necessary after the stringency wash. If storing blots, wrap them in Saran wrap so that they remain moist.

3.4. Signal Detection

Although fluorescein exhibits intrinsic fluorescence, the level of fluorescein on the blots is not sufficient to be detected by directly measuring the fluorescence of the labeled probe. Therefore, a detection module is used to visualize the colonies that contain hybridized fluorescein probes.

1. Incubate the blots, DNA-side up, for 1 h at room temperature in a mixture of 67.5 mL buffer A and 7.5 mL of Liquid Blocking agent. Shake gently at 75 rpm on an orbital shaker (*see* **Note 21**).
2. Transfer the blots, DNA-side up, to 27 mL of buffer A containing 0.5% BSA (w/v) and 5 μL of the anti-fluorescein-AP conjugate. Shake gently at 75 rpm for 1 h.
3. Wash the blots, DNA-side up, three times each for 10 min in 120 mL of buffer A containing 0.3% Tween-20 (w/v).
4. Drain off excess buffer and place the blot, DNA-side up, on a clean piece of Saran wrap on a flat surface.
5. Cover the blot evenly with CDP-Star detection reagent, ca. 3 mL per blot, and incubate for 5 min at room temperature.
6. Drain off excess detection reagent (*see* **Note 22**).
7. Place the blot in a detection bag provided with the CDP-Star reagent.
8. Place the blot into the digital imager, DNA-side up, and measure chemiluminescence using highest sensitivity with a clear filter. Vary the acquisition time to obtain a strong signal (*see* **Note 23**).

3.5. Data Analysis

3.5.1. Data Conversion

The acquired images are easily analyzed. First, the density of the detected signals of the array must be quantified. Then, the data of all sets of probes (binding to different genes but to the same region) must be compiled to determine the composition of each gene. A sample spreadsheet that facilitates this analysis can be found on the web at http://cheme.caltech.edu/groups/fha/probes/crossovers.html.

1. Quantify the signal intensity of each spot using appropriate software.
2. Export these data in an ASCII format which can be imported into an excel spreadsheet (Image-conversion.xls, sheet 1).
3. Define a signal intensity threshold using your controls to define the background signal, i.e., the signal from binding non-specifically to the negative controls.

These intensities will probably differ from blot to blot so a threshold must be defined for each. Intensities below this value are considered negative (false, = 0 on sheet 1) while intensities above are considered positive (true, = 2 on sheet 1). Intensities in between are also accounted for (intermediate, = 1 on sheet 1).

4. Data for one position are then pasted into another spreadsheet (Image-conversion.xls, sheet 2) to determine which parent occurred at this position. A "0" marks a position with no observed signal; an "X" marks a position with ambiguous results (at least two positive signals) (*see* **Note 24**).
5. These data are then combined in sheet 3 to determine the parent sequence present at each probed position for each clone.

3.5.2. Data Analysis

The probe hybridization method quickly generates a large amount of data that is not useful in its crude form. However, a great deal of information about the composition of the library, e.g., the relative amount of each parent in the library and the percent incorporation of each parent at each probe position can be extracted by simple spreadsheet analysis. This can reveal important biases in the library that can limit the diversity available for screening *(1)*.

In the context of assessing library diversity, the average number of crossovers (n_c) is an important parameter, since the diversity will scale with the number of combinations of "good" crossover sites taken n_c at a time. In our experience, 79% of crossovers occur in regions of identity longer than 5 bp, and 62% occur in identical regions longer than 10 bp *(1)*. Estimating n_c from probe hybridization data is complicated by the fact that two or more crossovers can be hidden between two probe sites, and thus just counting the average number of times that probe site *X* is occupied by one parent and site *X*+1 is occupied by another will underestimate the actual number of crossovers.

This problem can be circumvented using our method for calculating the average number of crossovers occurring between two probe sites given the set of probabilities that parent *a* is present at site *X* and parent *b* is present at site *X*+1. The method assumes crossovers are uniformly distributed between probe sites; this is generally a good assumption for probes separated by at least 150 bp. If crossovers are uniformly distributed, then there is a set of probabilities that a crossover will occur to parent *b* at nucleotide *x* given parent *a* at nucleotide *x*–1. We can write an equation that correlates these nucleotide-level probabilities to the probe-level probabilities that parent *a* is present at site *X* and parent *b* is present at site *X*+1. This equation can be iterated to solve for the nucleotide-level probability matrix and then we can extract the average number of crossovers that occur between any two probes for any two parents. The development of this method is summarized on our website at http://cheme.caltech.edu/groups/fha/probes/crossovers.html. Matlab code for implementing the method is included.

4. Notes

1. A kit containing fluorescein-11-dUTP, Terminal transferase, and Liquid Blocking reagent is available from Amersham-Pharmacia Biotech (Gene Images 3'-oligolabelling module, Cat. No. RPN 5770). This kit contains sufficient material for performing approx 70 blots of 7.5 × 11.5 cm. In addition, Enzo biochemicals (Farmingdale, NY) carries a similar fluorescein 3'-OH terminal labeling kit (Cat. No. 42631) that can be substituted when the Gene Images kit is not available.
2. Buffer A should be divided into 500-mL aliquots and autoclaved before use. In addition, once a flask is opened it should not be reused except on the same day. This reduces background problems during imaging that arise from bacterial contamination.
3. A kit containing anti-fluorescein antibody conjugated to alkaline phosphatase, liquid blocking reagent, CDP-Star detection reagent, and detection bags is available from Amersham-Pharmacia Biotech (Cat. No. RPN 3510). This kit contains sufficient material for 30 blots of 7.5 × 11.5 cm.
4. Many digital-imaging systems can be used for analyzing the chemiluminescence of the blot provided that array-analysis software is available. The software provided with the Biorad FluorS MultiImager can extract arrays from images, quantitate the signal, and report this data in exportable spreadsheet form.
5. To accurately determine the number of crossovers and estimate the relative likelihood of crossovers in several gene segments, the number of probes used should be given careful consideration. As the number of crossovers between two probe sites increases, the probe data will change less and less due to saturation effects. For a library containing a high crossover frequency, the probe data will only be useful to ascribe a lower bound to the number of crossovers. For a library constructed using 2–3 parents, the actual number of crossovers should be less than ~1.25 between neighboring probes when ~300 clones are considered *(1)*. Thus, the number of probes used should be slightly less than the expected number of crossovers. When more parents are used, fewer probes are required (e.g., one probe site per ~1.5 crossovers for >5 parents). The probes should be evenly spaced across the gene.
6. Choosing probes that bind to regions of low identity between the parental genes greatly simplifies the optimization steps during hybridization. When probes are used that exhibit low identity the stringency wash steps during hybridization (**Subheading 3.3.2.**) may not be required. The difference in T_m between a perfectly matched and a partially mismatched target/probe pair not only depends on the number of mismatches but also on the position of the mismatch, the type of mismatch, and the base pairs surrounding the mismatch. Under optimal conditions and with mismatches that significantly decrease T_m, it is theoretically possible to detect one to two mismatches. However, to facilitate the procedure probes should contain three or more mismatches *(1)*. If it is not be possible to design probes that bind to identical regions of the parental genes owing to too-high identity, then shift them a few bases up or downstream.
7. The melting temperatures (T_m) of probes can be easily calculated using **Equation 1**, where n is the number of nucleotides *(3)*. A more accurate prediction of

T_m can be obtained using the program Meltcalc (www.meltcalc.com) which takes into account nearest-neighbor thermodynamic data *(4)*. This software allows for easy calculation of the differences in T_m of probes exhibiting high sequence identity, i.e., those that differ in sequence at only a few base pairs. The recommended T_m of 62°C was calculated using Meltcalc with its default concentrations of 300 m*M* salt, 0 m*M* DMSO, and 0.1 μ*M* oligonucleotide probe.

$$T_m = 81.5°C + 16.6 \{\log_{10}(Na^+) + 0.41[\%(G+C) - (500/n)]\}$$

8. The National Center for Biotechnology Information has an alignment tool that can be used to search your vector for oligonucleotide binding sites (www.ncbi.nlm.nih.gov/blast/bl2seq/bl2.html) *(5)*. In addition, Amplify is an excellent program for simulating annealing of probes for PCR or hybridization (www.wisc.edu/genetics/CATG/amplify/) *(6)*.
9. Fluorescein is light-sensitive and therefore should be stored in the dark.
10. Most *Escherichia coli* strains can be used to array the library, but we recommend using BL21 cells, if possible, because this strain typically yields uniform cell growth.
11. Either LB- or M9-agar plates can be used for growing arrayed colonies on nylon membranes. However, M9 may be preferred for some bacterial strains, e.g., BL21 to slow colony growth and ensure uniformly-sized colonies on the array.
12. Using a picking robot to inoculate 384-well plate cultures results in more uniform culture growth than manual picking. Large differences cell density affect the quality of the DNA arrayed on the membrane. Dense cultures lead to sporadic large colonies which can overlap with other colonies arrayed on the nylon membrane.
13. Evaporation in wells near the edge of the 384-well plate will occur if the plate is not wrapped in Saran wrap. Significant evaporation can lead to non-uniform colonies.
14. Agar plates containing 2% (w/v) agar are used to prevent colony bleeding when stamping the membranes with the 384-pin replicator. Since bleeding of colonies can occur if the agar plates have excess condensation on the surface, the plates should be dried prior to use.
15. Use a few more membranes than you have probes in case of uneven growth or mistakes that could occur during processing of the membrane. This replicating step is the crucial step of the experiment and can be difficult to optimize. To ensure equal growth of colonies, place the replicating tool into the 384-well plate and vertically pull it out again; do not bump into the sides of the wells when lifting up. Place the stamp on the membrane, apply short but relatively strong pressure without moving horizontally, and lift vertically. Excessive pressure can lead to colony bleeding, but insufficient pressure can result in some pins not touching the membrane.
16. The incubation time depends on the bacterial strain and the media used. Growth should be stopped once the colonies can be barely seen; large colonies can lead to smearing of DNA during the lysing and denaturing step (**Subheading 3.2.2.**). A broad range of incubation times is required, e.g., 10–24 h, depending on strain, medium and insert. The optimal time must therefore be determined.

17. Sufficient denaturing (or neutralizing) solution should be used so that the membrane becomes completely wetted, but excess solution should be avoided so that the colony side of the membrane remains free of excess liquid. Wetting the colony side of the membrane will result in colonies running into each other and will yield diffuse, weak, and streaky hybridization signals *(7)*. Usually ~3–4 mL of solution is sufficient per membrane.
18. While the denaturing step lyses the bacterial cells, chunks of cell debris often remain on the surface of the membrane. These can be removed easily by wiping the colony side of the membrane with gloved fingers while it is submerged in the 2X SSC.
19. When optimizing hybridizations, initial experiments should be performed at a temperature that is 5°C below the T_m, and no stringency wash should be performed. Lower temperatures can be used when probes differ significantly. If no signal is initially observed, then hybridizations should be performed for a longer period of time over a range of temperatures until a good signal is observed. Occasionally a probe will not yield a strong signal even after optimization; we recommend not using these probes and identifying new probes proximal to the original ones that give a good signal. After a good signal is established, the stringency washes should be optimized (*see* **Note 20**).
20. Typically a stringency wash is required to reduce non-specific binding when oligonucleotide probes are directed towards regions of the genes that exhibit high identity. Stringency washes, when performed properly, result in a decrease in background without significant effect on the desired signal. However, if excessive stringency washes are used, then both the background and the signal can be removed. Therefore, this step requires careful optimization. Two variables affect the removal of background during the stringency wash, the salt concentration and the temperature at which the wash is performed. The lower the concentration of SSC buffer, the greater the stringency, i.e., more background is removed. Increasing the temperature also increases the stringency. To simplify optimization of this step, we recommend using a single temperature that is 10°C lower than the probe T_m and varying the SSC concentration. For low stringency washes, use 2X SSC and 0.1% (w/v) SDS, medium stringency washes use 1X SSC and 0.1% (w/v) SDS, and high stringency washes use 0.1X SSC and 0.1% (w/v) SDS.
21. Containers used for signal detection should be wiped with ethanol prior to use to prevent bacterial alkaline phosphatase contamination.
22. Excess detection reagent can result in a high overall background. To remove reagent from blot, lift one side of blot with tweezers until only one corner is touching the Saran wrap.
23. Excellent hybridizations typically require 1–2 min of signal detection. However, the detection system will produce chemiluminescence for >30 min. Therefore, if the signal is weak, longer acquisitions can be performed.
24. For the libraries analyzed in our study of chimeric dioxygenases *(1)*, 96.3% of the positions gave unambiguous results, 2.0% were ambiguous (X), and 1.2% gave no result (0).

References

1. Joern, J. M. , Meinhold, P., and Arnold, F. H. (2002) Analysis of shuffled gene libraries. *J. Mol. Biol.* **316,** 643–656.
2. Abecassis, V., Pompon, D., and Truan, G. (2000) High efficiency family shuffling based on multi-step PCR and in vivo DNA recombination in yeast: statistical and functional analysis of a combinatorial library between human cytochrome P450 1A1 and 1A2. *Nucleic Acids Res.* **28,** E88.
3. Sambrook, J. and Russell, D. W. (2001) *Molecular Cloning, A Laboratory Manual*, Cold Spring Harbor Laboratory Press, Cold Spring Harbor, NY.
4. Schutz, E. and von Ahsen, N. (1999) Spreadsheet software for thermodynamic melting point prediction of oligonucleotide hybridization with and without mismatches. *Biotechniques* **27,** 1218–1224.
5. Tatusova, T. A. and Madden, T. L. (1999) BLAST 2 Sequences, a new tool for comparing protein and nucleotide sequences. *FEMS Microbiol. Lett.* **174,** 247–250.
6. Engels, W. R. (1993) Contributing software to the internet: the Amplify program. *Trends Biochem. Sci.* **18,** 448–450.
7. Darling, D. C. and Brickell, P. M. (1994) *Nucleic Acid Blotting*, Oxford University Press, New York, NY.

21

Sequence Mapping of Combinatorial Libraries on Macro- or Microarrays

Experimental Design of DNA Arrays

Valérie Abécassis, Gilles Truan, Loïc Jaffrelo, and Denis Pompon

1. Introduction

Sequence mapping consists in the characterization of a set of sequence segments in a library of related genes or PCR products. This procedure can be applied to analyze natural sequence variants, for example allelic distribution of human genomic sequences for disease, drug metabolism prediction, or sequence associations in combinatorial libraries of genes used in directed evolution. Sequence mapping of combinatorial libraries have a lot of potential interests including diversity characterization, bias and artifact analysis, library improvement by robotic equalization, and population-wide structure-function analysis *(1,2)*. Sequence mapping differs from classical sequencing by the fact that prior knowledge of possible sequence alleles is generally requested. Nevertheless, at the limit of the technique, true resequencing can be performed with single-base resolution. A second advantage of sequence mapping over sequencing is that the former is expected to have high-throughput and low-cost capabilities suitable for routine applications with libraries, whereas the latter is more dedicated to characterizing individual clones. Our goal was to develop low-cost, sensitive, and rapid sequence mapping methods. The first method describes a dot-blot based approach and is adapted to libraries containing less than 1000 clones. The second one, more sensitive, is microarray based and adaptable to larger library characterization. The sequence characterization of a combinatorial library of mosaic P450 in pYeDP60 *(3)* built from the human *CYP1A* family members (*CYP1A1* and *CYP1A2*) is used to illustrate the techniques.

From: *Methods in Molecular Biology, vol. 231: Directed Evolution Library Creation: Methods and Protocols*
Edited by: F. H. Arnold and G. Georgiou © Humana Press Inc., Totowa, NJ

2. Materials

2.1. Material in Common for Macro- and Microarray Methods

1. *E. coli* strain DH5-1.
2. Bacterial medium TB or Luria Bertani Broth (LB) ***(4)***.
3. Ampicillin (stock solution at 100 mg/mL, store at –20°C), in LBA, ampicillin is used at 100 µg/mL.
4. 384-well microtiter plate (Genetix, Queensway, UK).

2.2. Specific Materials for Macroarrays

1. Square LBA plate (25 × 25 cm).
2. 384-well replicator (Genetix, Queensway, UK).
3. N+ nylon filters (Amersham, Buckinghamshire, UK).
4. Standard Southern blot hybridization and washing equipment.
5. γ-ATP^{32} (4000 Ci/mmol).
6. T4-polynucleotide kinase (New England Biolabs, Beverly, MA).
7. Autoradiography films (Kodak, Rochester, NY).
8. 20X sodium citrate/sodium chloride (SSC), pH 7.0.
9. 20X SSPE.
10. 100X Denhardt's solution: 5 g Ficoll 400, 5 g polyvinylpyrolidone, 5 g bovine serum albumin (BSA), completed to 250 mL with pure water (conserved at –20°C to prevent micro-organism contamination) ***(4)***.
11. Bacterial cell lysing solution: sodium dodecylsulfate (SDS) 0.2%.
12. DNA denaturation and fixation solution: 0.5 *M* NaOH, 1.5 *M* NaCl.
13. Neutralization solution: 1 *M* NaCl, 50 m*M* Tris-HCl, pH 7.0.
14. Washing solution : 2X SSC.
15. Hybridization buffer : 5X SSPE, 5X Denhardt's solution, 0.5% SDS.
16. Washing buffer: 2X SSC/0.1% SDS.

2.3. Specific Materials for Microarrays

1. Oligonucleotide primers for PCR amplification of the shuffled genes. For pYeDP60, we used: 5': CGTGTATATAGCGTGGATGGCCAG and 3': GAA GCACCACCACCAGTAGAG).
2. *Taq* DNA polymerase (Stratagene, La Jolla, CA).
3. PCR 96-well plates (Corning, Acton, MA).
4. Multiscreen system for PCR product purification (Millipore, Bedford, MA).
5. UV transparent microtiter plates (Greiner, Kremsmünster, Austria) and Power Wave X spectrophotometer (Bio-Tek Instruments Inc., Winooski, VT) to measure DNA concentration. Alternatively, an estimation can be performed using agarose electrophoresis.
6. Vacuum DNA concentrator.
7. Cy3- or Cy5-labelled oligonucleotides.
8. Poly-L-lysine (LaboModerne, Paris, France) or amino-silane (Sigma or Corning) glass slides.
9. SYBR GREEN II solution (Molecular Probes, Eugene, OR).

10. GeneTac Hyb4 automated hybridization station (Genomic Solutions, Ann Arbor, MI) or any equivalent hybridization station.
11. GenePix scanner (Axon, Union City, CA) or equivalent.
12. GenePix software (Axon) for hybridization data reading and processing or equivalent.

3. Methods

Two closely-related approaches for sequence mapping are described. The first one is based on hybridization of a single short oligoprobe per locus to test on plasmids from bacterial lysates or colonies. This approach is adapted to characterization of small libraries and offers a good quality/cost ratio but limited reliability for discrimination of closely-related sequences. The second approach is based on an original DNA chip strategy allowing characterization of large libraries with excellent reliability, but requires significant hardware.

3.1. Probe Design

Figure 1 illustrates the strategy used to design probes. For both methods, oligonucleotides were preferentially chosen in the regions of lower sequence similarity between the two parental sequences. Position of mismatches in the oligonucleotides is also of importance. Mismatches in the central regions of oligonucleotides are more discriminant than ones at the edges. The length of oligonucleotides is less critical in the macroarray approach, as hybridization of individual probes can be performed in optimized stringency conditions (salt, temperature, etc.). Nonetheless, probes longer than 18 bases and shorter than 23 bases were preferred (**Table 1**). The melting temperatures of those probes is not a too-critical point either, as each hybridization can be conducted separately. Use of probes that alternatively target the two parental types along the sequence offers a significant advantage for optimal data analysis (*see* **Fig. 1**). Oligonucleotide design is much more critical for microarray analysis as competitive hybridization is conducted. Six pairs of oligonucleotide probes were designed to 1) match the same sequence segments as the ones used for macroarray analysis and 2) match the melting temperatures in each pair (calculated using the DNAMAN software, Lynnon Biosoft, Vaudreuil-Dorion, Canada) (**Table 2**). Labeled oligonucleotides were synthesized by ESGS (Cybergene, Evry, France). The *CYP1A1* and *CYP1A2* probes were labeled with Cy3- and Cy5-dyes, respectively. The two probes belonging to the same pair were simultaneously hybridized to the same slide.

3.2. Macroarray Preparation

The method described below includes 1) preparation of the Nylon filters preparation, 2) bacterial cell lysis and DNA fixation, and 3) the hybridization and washing conditions.

Fig. 1. Probe design strategy. (**A**) macroarrays; (**B**) microarrays.

Table 1
Sequences of Oligonucleotide Probes Used for Macroarrays

Probes	5'→3' sequence	Length	Start positions
Probe 1 (*CYP1A2*)	GCATTGTCCCAGTCTGTTCCCTTC	24	3
Probe 2 (*CYP1A1*)	CCGGCGCTATGACCACAACCACCAAGAACTG	31	612
Probe 3 (*CYP1A2*)	AGACTGCCTCCTCCGGGAACCCCC	24	683
Probe 4 (*CYP1A1*)	GCTGGATGAGAACGCCAATGTC	22	879
Probe 5 (*CYP1A2*)	CGGGGAAGTCCTGGCAAGTGG	21	1377
Probe 6 (*CYP1A1*)	CACTTCCAAATGCAGCTGCGCTCT	24	1513

3.2.1. Filter Preparation

1. Transform electrocompetent DH5-1 bacterial cells by electroporation using standard molecular biology methods *(4)* and plate onto LBA solid medium.
2. Select isolated bacterial colonies from the library and inoculate with a toothpick 378 wells of a 384-well plate containing 40 µL of TB liquid medium containing 100 µg/mL ampicillin. The six remaining wells are used as controls and are inoculated with DH5-1 cells previously transformed with p1A1/V60 or p1A2/V60.
3. Grow the micro-cultures for 24 h at 37°C without agitation, and then replicate using a 384-well microtiter plate replicator onto six Nylon N+ membranes. It is important to shake the plate before replication as the cells sit in the bottom of the plate.

Table 2
Characteristics of Oligonucleotide Probes Used for Microarrays

Probes	5'→3' sequence	T_m (°C)	Length	Start positions
Probe set 1 (*CYP1A1*)	GCTTTTCCCAATCTCCATGTCGG	68	23	3
Probe set 1 (*CYP1A2*)	GCATTGTCCCAGTCTGTTCCCTTC	67	24	4
Probe set 2 (*CYP1A1*)	CGCTATGACCACAACCACCAAGAAC	68	25	616
Probe set 2 (*CYP1A2*)	GACAGCACTTCCCTGAGAGTAGCGAT	67	26	617
Probe set 3 (*CYP1A1*)	GTGGTTGGCTCTGGAAACCCAG	69	22	679
Probe set 3 (*CYP1A2*)	AGACTGCCTCCTCCGGGAACC	68	21	683
Probe set 4 (*CYP1A1*)	GCAGCTGGATGAGAACGCCAAT	68	21	876
Probe set 4 (*CYP1A2*)	TAGAGCCAGCGGCAACCTCATC	68	21	885
Probe set 5 (*CYP1A1*)	CGGTGAGACCATTGCCCGC	70	19	1374
Probe set 5 (*CYP1A2*)	CGGGGAAGTCCTGGCCAAGTGG	69	21	1377
Probe set 6 (*CYP1A1*)	GAGCACTTCCAAATGCAGCTGCG	70	21	1510
Probe set 6 (*CYP1A2*)	GAACATGTCCAGGCGCGGC	70	19	1513

4. Place each filter at 37°C for 12 h onto the square plate containing LBA solid medium.

3.2.2. Bacterial Cell Lysis and DNA Binding

1. Saturate four Whatmann 3MM pieces of paper with four different solutions (10% SDS; 0.5 *M* NaOH, 1.5 *M* NaCl; 1 *M* NaCl; 50 m*M* Tris-HCl, pH 7.0; 2X SSC (the Whatmann sheets must be very wet to saturate the nylon filter).
2. Place the nylon filter 3 min on the first solution, and 5 min on each of the remaining solutions. Dry the filters for 30 min on a 3MM Whatmann paper. The filters can be conserved as such at 4°C for weeks.

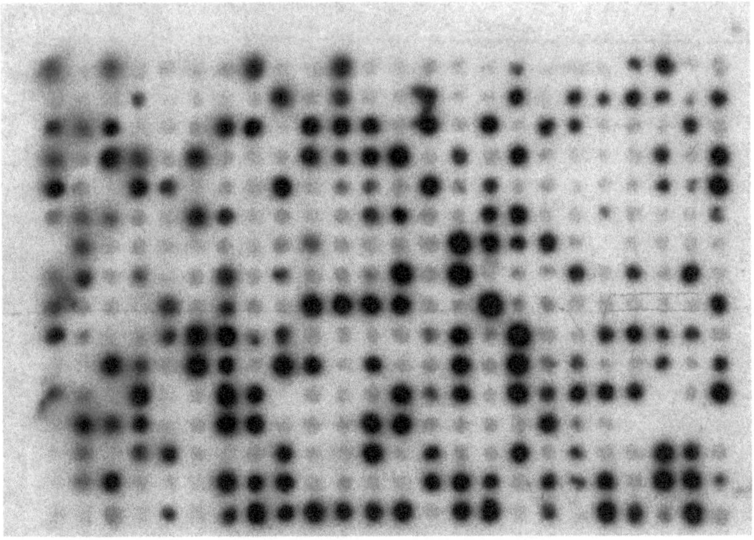

Fig. 2. Example of an autoradiography generated with a macroarray.

3.2.3. Hybridization and Washing

1. Place the filters for 5 min onto a 3MM Whatmann paper soaked with 2X SSC.
2. Denature herring sperm DNA (1 mg/mL) at 100°C for 5 min and cool it immediately on ice. This non-homologous DNA is used to saturate the nylon membrane.
3. Pre-hybridize the filters for 1–3 hours at 65°C with 25 mL of the hybridization buffer containing 0.5 mL of the denatured DNA solution.
4. Labeling of the oligonucleotides can be performed during pre-hybridization. Incubate 11 pmoles of each probe for 2 h at room temperature with 3.2 pmoles of γ-ATP^{32}, 20 U of polynucleotide kinase, and 18 μL of polynucleotide kinase buffer following manufacturer's recommendations.
5. Add each probe to the pre-hybridization solution and low the temperature to 42°C. Hybridize 2–12 h, depending on the probe (*see* **Note 1**).
6. Wash the filters at room temperature (or suitable higher temperature to increase stringency) in 2X SSC containing 0.1% SDS for 10 min and autoradiograph for 3 h (*see* **Note 2** and **Fig. 2**).

3.2.4. Results Analysis

Analysis of the results is described in the chapter on bioinformatic treatment of data.

3.3. Microarrays Preparation

The method described below includes 1) preparation of the target DNA, 2) spotting conditions, and 3) hybridization conditions.

3.3.1. Preparation of the Target DNA

3.3.1.1. PCR Amplification

1. Select 381 independent bacterial colonies from the library and inoculate a 384-well plate containing 40 μL of LBA liquid medium. The three remaining wells are inoculated with DH5-1 cells previously transformed with p1A1/V60, p1A2/V60, and pYeDP60.
2. After 24 h growth, four 96-well plates PCR amplifications are performed in a 50 μL PCR reaction using 1) 0.5 U of *Taq* polymerase, 2) 1 μL of each microculture, and 3) primers located in the promoter and terminator regions of pYeDP60 *(2)*.
3. The PCR program used is: 1 cycle of 94°C for 3 min; 30 cycles of: 1 min at 94°C, 1 min at 56°C for and 3 min at 72°C; and finally a 7 min step at 72°C.
4. Check the quality of the PCR amplification by agarose gel electrophoresis of 20 randomly chosen samples.

3.3.1.2. PCR Product Purification

1. Add 40 μL of water to the PCR mix and purify the amplified DNAs with the Multiscreen 384-well filter plate using the manufacturer's protocol and at least one wash step.
2. Add 40 μL of water in each well and agitate to resuspend the PCR products.
3. DNA concentrations can be estimated using agarose gel electrophoresis or measured with a Power Wave X spectrophotometer and UV-transparent plates.
4. Transfer the DNA into a standard 384-well plate and concentrate by evaporation under vacuum.
5. Resuspend the DNA with a spotting solution containing 50% high purity formamide in water (*see* **Note 3**) to a final concentration higher than 100 ng/μL *(5)*.

3.3.2. Spotting Conditions

1. Spot DNA samples with poly-L-lysine (or amino-silane) coated slides (75 × 25 mm) and a GeneTac G3 robot (Genomic Solutions) with pins of 150 or 200 μm. DNA samples are printed in quadruplicate with a center-to-center spacing of 300 μm.
2. Dry the slides at room temperature and the cross-link the DNA with a UV lamp for 5 min at 250,000 μJ/cm^2. Slides can be kept in the dark at room temperature for weeks.

3.3.3. Hybridization and Washing

1. Filter all the solutions with 0.2-μm Corning filters. Pre-hybridization, hybridization and washing steps are performed in the automated hybridization station.
2. Just before use, place the slides for 5 min in 2X SSC; 0.1% SDS, for 3 min in 0.2X SSC, for 2 min in water, and then dry in a speed-vac for 5 min.
3. Submit the slides to a 30 min pre-hybridization step at 42°C in 5X SSC, 0.2% SDS; 5X Denhardt's.

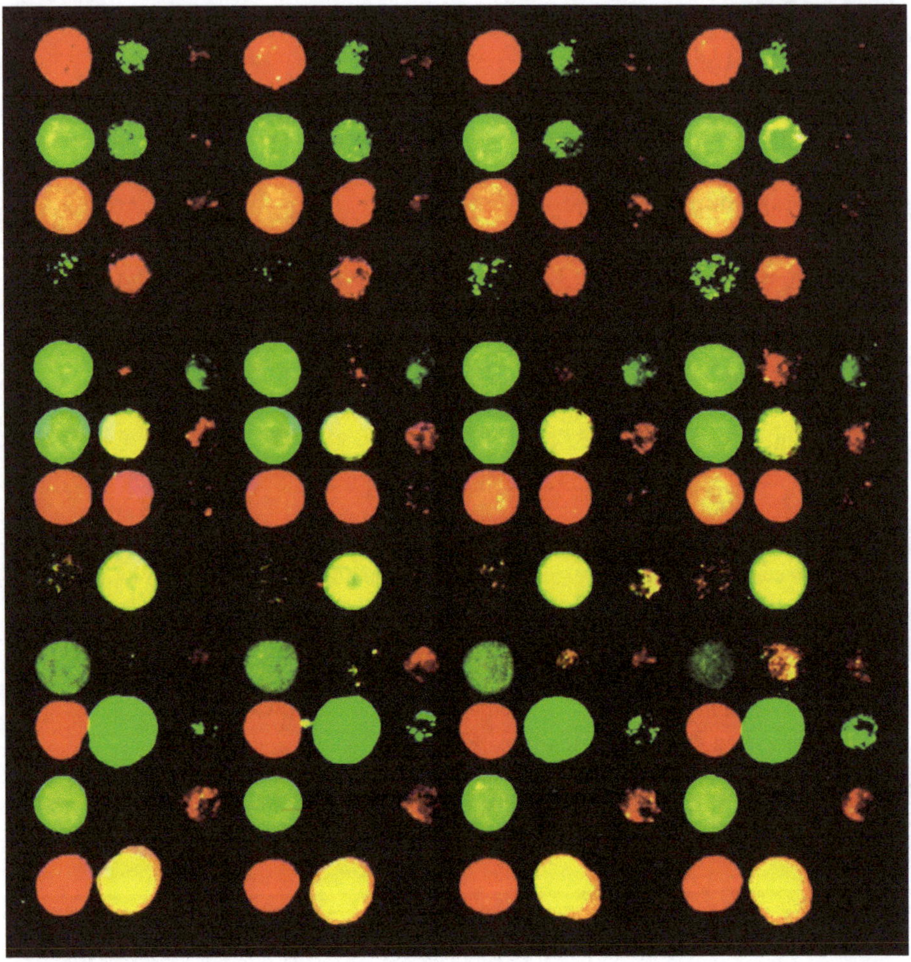

Fig. 3. Example of a scan of one slide generated by the microarray strategy.

4. Perform the hybridization step at 42°C for 1.5 h in the same solution containing 1–5 picomoles of each probe pair.
5. Wash the slides two times (30 s each) with the following solutions: 2X SSC/0.2% SDS (at 25°C), 0.2X SSC (at 35°C), and 0.05X SSC (at 25°C).
6. Rinse immediately in water (no more than a few seconds) and dry under vacuum for 5 min.

3.3.4. Data Analysis

Slides are introduced in the GenePix 4000B scanner and scanned at 532 and 635 nm (*see* **Fig. 3** for a scan example). GenePix software is used for primary data analysis.

4. Notes

1. To ensure reproducibility of results between filters, at least two filters were hybridized with each probe.
2. Depending on the background obtained, washing can be performed in various stringency conditions. Of course the autoradiography time has to be adjusted to the radioactivity present on the filters.
3. Usually formamide was deionized with resins before use. Unfortunately, we found that this step can release fluorescent products that enhanced background on the slides. To circumvent this problem, high purity formamide was purchased, aliquoted and stored at –20°C.
4. To improve the PCR product purification using the Multiscreen 384-well filter plate the microtiter plate must be washed two times with 60 µL of pure water. Between each washing step put the plate onto a microtiter plate shaker for 5–10 min at room temperature to ensure good resuspension of the samples. After the final purification, following the addition of 40 µL of water the micro-titer plate has to be shaken for 5 min at room temperature.
5. Another important point during operation is to avoid uncontrolled water transfer through the plate membrane by placing a wet piece of Whatmann paper under the plate.
6. To estimate the spotting quality, treat one slide of each series by a 1:10,000 dilution in H_2O of SYBR GREEN II solution for 5 min followed by multiple washes in double-distilled water before scanning at 532 nm.

Acknowledgments

The work was supported in part by a grant from the GIP Hoechst Marion Roussel. Valérie Abécassis is a pre-doctoral fellow supported by a fellowship from the Ministère de la Recherche et de l'Enseignement Supérieur. The microarrays were generated in the Gif-Orsay DNA array facility with the help of Drs. Lawrence T. Aggerbeck and David Rickman.

References

1. Joern, J. M., Meinhold, P., and Arnold, F. H. (2002) Analysis of shuffled gene libraries. *J. Mol. Biol.* **316,** 643–656.
2. Abecassis, V., Pompon, D., and Truan, G. (2000) High efficiency family shuffling based on multi-step PCR and in vivo DNA recombination in yeast: statistical and functional analysis of a combinatorial library between human cytochrome P450 1A1 and 1A2. *Nucleic Acids Res.* **28,** E88.
3. Urban, P., Cullin, C., and Pompon, D. (1990) Maximizing the expression of mammalian cytochrome P-450 monooxygenase activities in yeast cells. *Biochimie* **72,** 463–472.
4. Sambrook, J., Fritsch, E. F., and Maniatis, T. (1989) *Molecular Cloning, A Laboratory Manual, 2nd ed.,* Cold Spring Harbor, Laboratory Press, Cold Spring Harbor, NY.

5. Diehl, F., Grahlmann, S., Beier, M., and Hoheisel, J. D. (2001) Manufacturing DNA microarrays of high spot homogeneity and reduced background signal. *Nucleic Acids Res.* **29,** E38.

22

Sequence Mapping of Combinatorial Libraries on Macro- and Microarrays

Bioinformatic Treatment of Data

Denis Pompon, Gilles Truan, and Valérie Abécassis

1. Introduction

Sequence mapping of combinatorial libraries is of interest for evaluation of library diversity and homogeneity as well as for bias detection and analysis. This is particularly useful for library improvement by robotic equalization. Macro- and microarray based experimental procedures adapted for this purpose are described in the previous chapters. This chapter aims to delineate procedures and tricks required for data interpretation. Incorrect interpretation of hybridization signals usually occurs because of the highly parallel data treatment using global and automated discrimination criteria when experimental data by themselves are subject to local perturbations and artifacts that would require individual spot examination. In addition, sequence changes involving a few or single base pairs correspond to very limited differences in probe hybridization properties and thus to a low signal-to-noise ratio. We describe here a method for macroarray analysis that is mainly semi-empirical but involves specific data analysis techniques for self-validation. The advantage is of course simplicity, but the drawback derives from subjective and user-dependent interpretations of relative signal intensities. In contrast, the method for interpretation of microarray data relies on statistical analysis and well-defined procedures and is not only giving robust sequence attribution, but also an evaluation of the reliability of the treatment. This method requires the use of a set of specific Excel® macros that are freely distributed. Finally, sequence signature calculation is described for both types of data. This signature constitutes a population-wide characterization of the sequence pattern of the combinatorial library.

From: *Methods in Molecular Biology, vol. 231: Directed Evolution Library Creation: Methods and Protocols*
Edited by: F. H. Arnold and G. Georgiou © Humana Press Inc., Totowa, NJ

2. Materials

1. Model Excel sheets including the suitable formulae and visual basic routines: "Macro_Process" for macroarray analysis, "Micro_Process" for microarray analysis, and "Sign_Process" for sequence signature analysis. These files (.xls in PC format) can be requested by mail to the authors. They are given as such without any form of guarantee and will generally need adaptation for specific purposes.
2. PC computer with at least 128 Mb memory.
3. Microsoft Excel® software (versions 97, 2000, or XP).

3. Methods

This section is divided into three parts: 1) the analysis of data from macroarrays, 2) the analysis of data from microarrays, and 3) the calculation of sequence signatures. It is practically impossible to design a "universal" bioinformatic tools. The described macros are models that can be easily adapted to specific purposes. Minimum skills in manipulating Excel® formulae and graphics as well as Visual Basic® programming are always useful and sometimes required.

3.1. Macroarray Data Analysis

This chapter is relevant for macroarray sequence mapping based on hybridization of a library spotted on Nylon membranes and hybridized with a single ^{32}P- or ^{33}P-labeled oligoprobe at a time. Nevertheless, in the case of using double radioactive labeling and a suitable multi-channel imager, the data will be better analyzed using the methods described later for microarray analysis. In the single probe per locus approach, the major difficulties arise from the rather poor reproducibility of macroarray hybridization signals owing to local fluctuations in membrane properties, dispersion of spotted DNA amounts, and variations in labeling, hybridising, and washing conditions. These difficulties can be partially overcome using local positive/negative/neutral discrimination criteria requiring visual examination and by comparative analysis of replicas. The following description applies to sequence mapping of libraries resulting from family shuffling between two parental sequences, but can be extended to a larger number of parents. Two alternate strategies can be used for probing. In the "quick" strategy, a single probe per sequence segment is used. The probes are chosen along the sequence to alternatively match parental types. Absence of signal is then interpreted as the opposite parental type at a given probe position. This approach is minimal in terms of cost and work and is considered safe after statistical validation when parental sequences are rather divergent. In the "full" strategy, two or more probes matching each possible parental type are used per position. This is useful to improve discrimination between more closely-related sequences but dramatically increases cost and time requirements.

3.1.1. Data Acquisition

Data acquisition can be performed using either autoradiography film or a radioactive imager like a Storm® or a Phosphoimager®. Quantitative signal analysis was not always found useful when using the single-probe strategy, as visual inspection was always required and semi-empirical attribution frequently preferred. The following procedure was used for data acquisition on libraries resulting from family shuffling of two parental sequences ***(1)***.

1. Hybridization intensities for control spots (parental sequences A and B and non-hybridizing sequences) are first examined to check signal discrimination. Hybridization and/or washing conditions are reinvestigated in the case of poor discrimination.
2. Spot values for each probe are classified into three groups (1, 2, 3) by visual inspection and comparison of the signal intensities with the ones for close neighbors and control spots. Group "2" was assigned to questionable signal intensities. Local comparisons must be privileged over evaluation of absolute intensities because of the frequent inhomogeneity between signals coming from different areas of the macroarray.
3. Readings are transferred into the input matrix of the Excel® "Raw Data" sheet. Ambiguous signals "2" can be subsequently reattributed either to parental type "1," "2," or discarded by activating software switches.

3.1.2. Data Validation

1. Proportions in the library of each parental allele at a given sequence position are calculated. Using the "quick" (single probe per segment) strategy, observation of rather similar or randomly fluctuating proportions of each parental type all over the sequence is a good indication for a clean attribution. In contrast, a regular oscillation of the calculated proportion matching the parental type choice for subsequent probes generally indicates erroneous allelic type attribution. Data from the "full" strategy are better interpreted using the "quick" strategy twice, keeping as valid the consistent attributions for the alternate probe sets and toggling contradictory attributions into group "2."
2. Analysis of the cross-correlation between parental type attributions for adjacent sequence segments is the second validation method. Actual DNA shuffling methods usually lead to limited probability of exchange between parental segments that are close in the primary sequence or separated by low similarity regions. Analysis of the cross-correlation patterns are thus a useful control to check sequence attribution quality as incorrect attribution statistically leads to a loss of cross-correlation.
3. Make sure that, in the "quick" strategy, at least one segment must be probed by two parental probes because of the frequent presence in libraries of cDNA free plasmids that can be interpreted as negative for all single-probe hybridizations and create an artifactual population of mosaic genes having a different parental type at each probe. Such phenomena are easily detected as a corresponding frequency peak in the sequence signature of the library.

3.2. Microarray Data Analysis

Microarrays differ from macroarray by the use of fluorescent double-labeling and higher redundancy. Two (Cy3- and Cy5-labeled) oligoprobes with overlapping targets and matching the possible allelic types are used for each probed segment. Experimental details about probe choice and experimental condition adjustments are given in a separate chapter. The very large number of spots to analyze (up to several thousand per slide) make visual inspection of individual spots for allele attribution impossible. Automation of this task is based on several steps: 1) data acquisition and flagging on a scanner, 2) data attribution and pre-processing, 3) adjustment of filter parameters, and 4) statistical modeling of the filtered population and allelic type attributions. These steps are performed with Excel® sheets according to the "Micro_Process" model. These sheets include suitable formulae and Visual Basic® routines. The only manual task consists of pasting scanner data and adjusting the program parameters using the built-in visual interface.

3.2.1. Data Acquisition and Flagging

The last experimental step of chip processing consists of the double-wavelength fluorescence scan of the hybridized slides. Details for this step are hardware- and software-specific, but involve common rules like choosing suitable laser powers and photo-multiplier gains to match, as well as possible green and red average signal intensities. A set of slides from a single library to be characterized are preferentially scanned using the same scanner settings. Each spot must be additionally toggled (generally automatically by the scanner software) as "good" or "bad" during the process. Use of a spotting geometry allowing quick visual inspection of "replica" signals is useful to facilitate flag editing when requested.

3.2.2. Data Attribution and Pre-processing

Data attribution consists of associating each spot on the slide to a microtiter plate well and finally to a clone in the library. This is a hardware- and software- (spotter and scanner) specific task. An example of automation for a Genomic Solution spotter/Axon scanner hardware is available on request, but such software needs to be adapted for different hardware configurations. Following this attribution step, medians of intensities minus background are calculated for the green and red channels and results transferred to the "Input" page of the "Micro_Process" program. This program can be fully operated from its parameter and graphic interface page (*see* **Figs. 1** and **2**). The optional pre-processing step includes signal corrections before calculation of discrimination indexes. The lower part of the graphics interface (*see* **Fig. 2**) includes X–Y plots of the raw and corrected signals. Using these plots, adjust the OFG and OFR offset parameters in order to center points for control (not hybridizing) DNA spots on the axis origin. Similarly, use the slope correction parameters (SCG and SCR) to realign, if necessary, the bottom and left limits of the cloud of points with the X and Y axis. These corrections are generally not required with a good data set. In such a case, set the four described parameters to zero (no correction). Discrimination indexes and corresponding plots are automatically generated at this point.

3.2.3. Adjustment of Filter Parameters

Filters aim to discriminate between spots for which sequence attribution can be safely performed, spots with noisy signals, and spots with non-interpretable signal. This last group generally corresponds to mixes of sequences or to the presence of unexpected alleles (*see* **ref. 2** for theoretical considerations). Proceed as follows:

1. On the graphic interface set selectors to the "raw signal" and "mode 1" positions and parameter NFC (number of frequency classes) to 40.
2. Set LST (low signal threshold) to 0, and adjust weighting factor WF (*see* **Fig. 3**) to a value for which frequency curve exhibits two (ideal case) or three (more frequent) clearly defined modes. This adjustment is only to facilitate visualization and will not affect subsequent calculations.
3. Shift selector to "low signal" and increase LTS up to the largest value for which the frequency curve behaves like a Gaussian-shaped and zero-centered distribution with little or no shoulders (*see* **Fig. 4**). This central distribution may be completely absent from good data set. In this case, set LST to zero.
4. Turn selector to "mixed sample" and "mode 2". Set DT (discrimination threshold) to 0.75 and MDT (minimal distance threshold) to three times the LST value.
5. Test values of DT ranging between 0.3 and 0.8, and MSD values between 2- and 10-times LST and keep the choice giving the best discrimination between "abnormal" and regular spots (*see* **Fig. 5**).
6. Turn selector to "cleaned data." Check, in the cumulative frequency curve, that the size of the population of "interpretable data" represents at least 80% of the total population. The frequency curve must be now clearly bimodal with modes close to +1 and –1 (*see* **Fig. 6**, top) indicating that sequence identification can be successfully performed. The negative values of indexes frequently present one or several shoulders on the frequency curve. This is not a major problem provided that little or no frequency is observed for index values in the [–0.2, +0.2] range. If this is not the case, try to increase LST parameter keeping MDT> LTS. If this fails, your experimental data may be of unadequate quality and has to be improved. If everything is OK, you are now ready for sequence attribution.

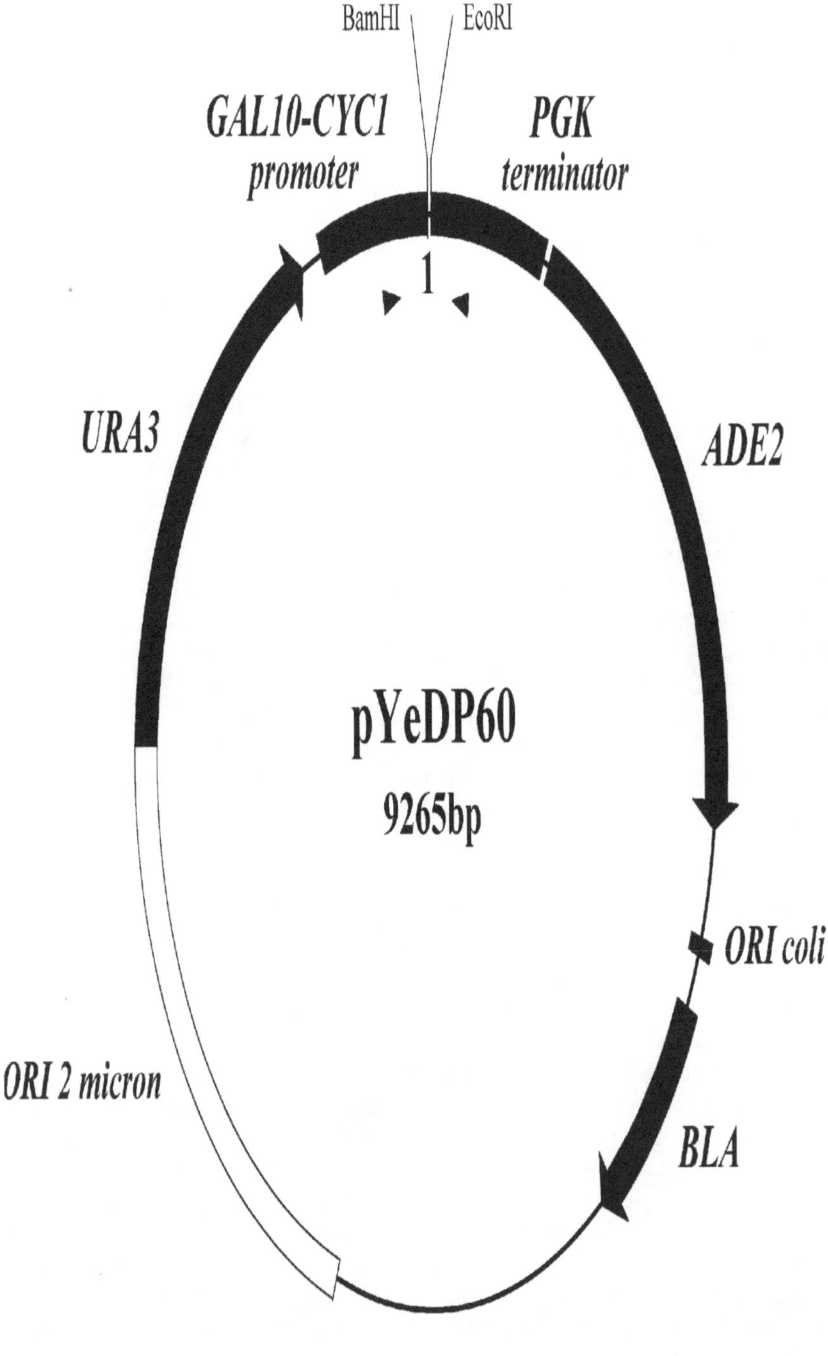

Fig. 1. Top panel of the graphics interface panel of the "Micro_Process" program following treatment of a typical sequence mapping microarray.

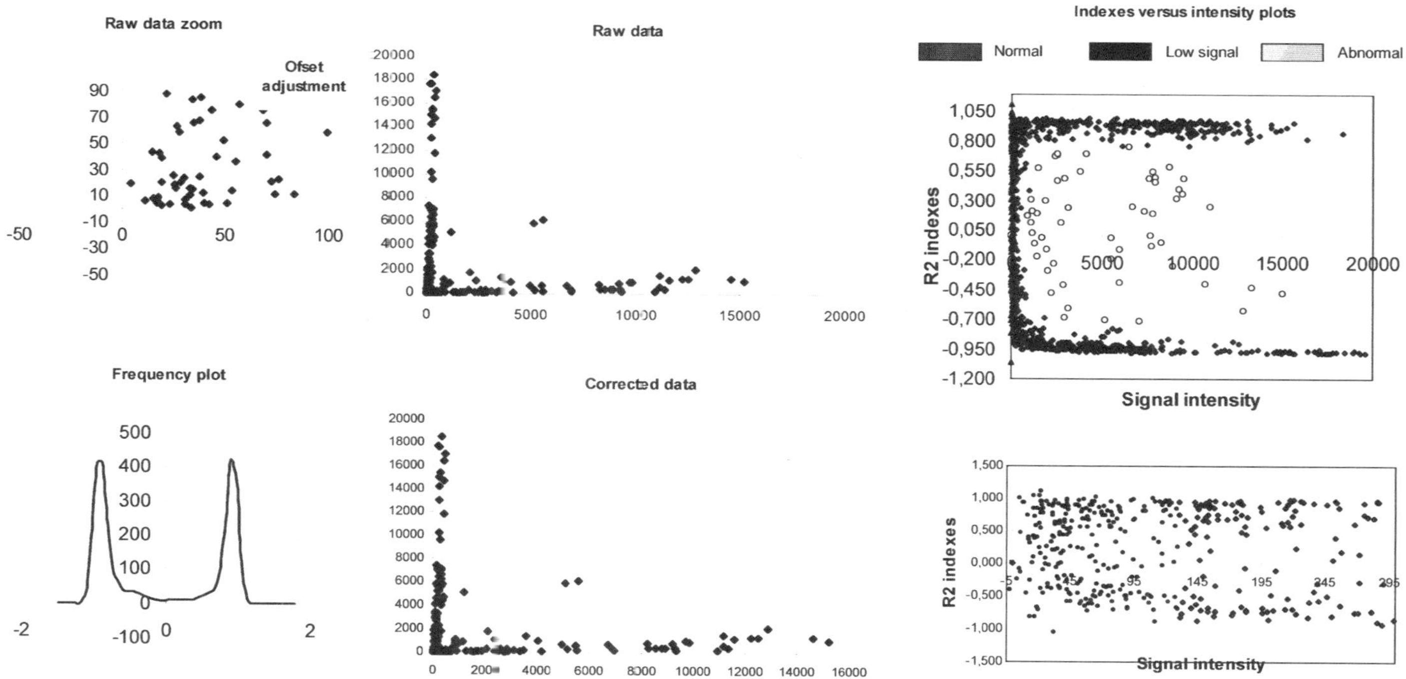

Fig. 2. Bottom panel of the graphics interface panel of the "Micro_Process" program following treatment of a typical sequence mapping microarray.

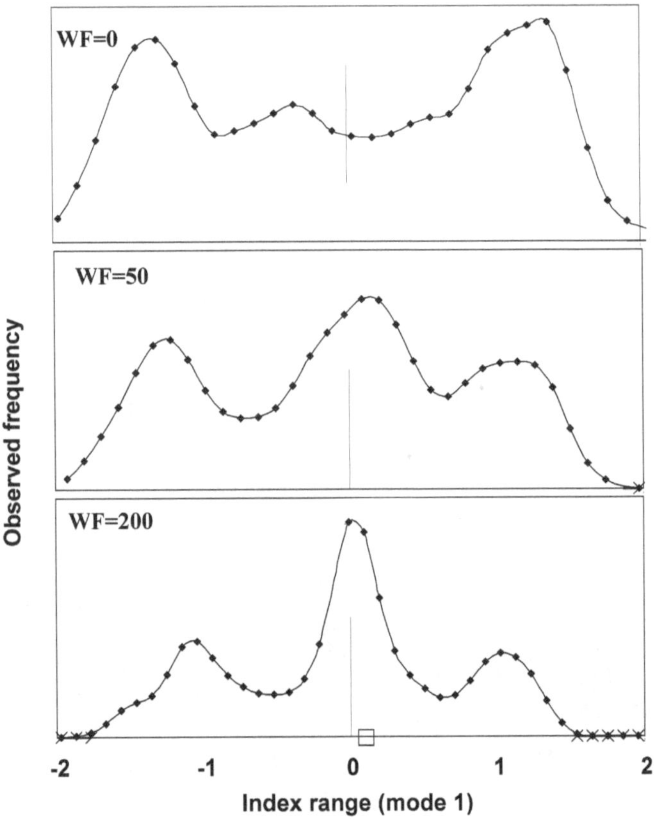

Fig. 3. Effect of the "weighing factor" (WF) parameter on the frequency of observation of the different classes of discrimination index values (mode 1, total spot population). Top, no weighting of ratios (WF = 0); center; partially adjusted parameter (WF = 50); bottom, optimally adjusted parameter (WF = 200).

3.2.4. Statistical Modeling of Population and Sequence Attribution

1. Count the number of visually detectable distributions on the frequency curve. This is ideally two, but three or four are frequent figures, and five is a practical limit. Set the GN parameter to this value.
2. Manually assign values (by trial and error) to parameter C1 (center), A1 (amplitude), and S1 (dispersion) to fit one of the frequency distribution with a Gaussian curve. Repeat for one or more of the other groups of parameters to fit each visible distribution (*see* **Fig. 6**, center) and to achieve the best fit of the experimental by the theoretical distributions on the top right panel of the visual interface (*see* **Fig. 1**).

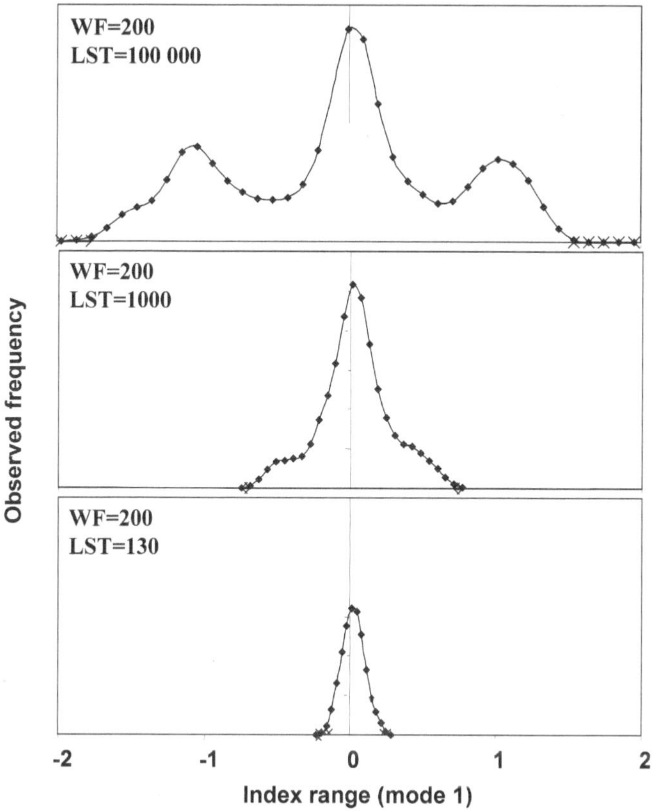

Fig. 4. Effect of the "low signal threshold" parameter (LST) on the frequency of observation of the different classes of discrimination index values (mode 1, total spot population). Top, no filtering (LST = 100,000); center; partially adjusted parameter (LST = 1000); bottom, optimally adjusted parameter (LST = 130).

Following such manual setting, launching the Excel® solver to minimize the software calculated fit error (FE value) may be helpful for improve parameter adjustment, but this can also led to erratic results.

3. Associate to each Gaussian curve a sequence allele type using the attached selectors that offer four possibilities "A" or "B" sequences alleles, "A+B" or "Ignore". This last option removes the corresponding population from the calculation, when "A+B" option will perform the "more probable" attribution. Finally the probability that the probed sequence segment belongs to one or the other of the tested allele is automatically calculated (left part panels of graphic interface) for each PCR product and can be used for further calculation or action.

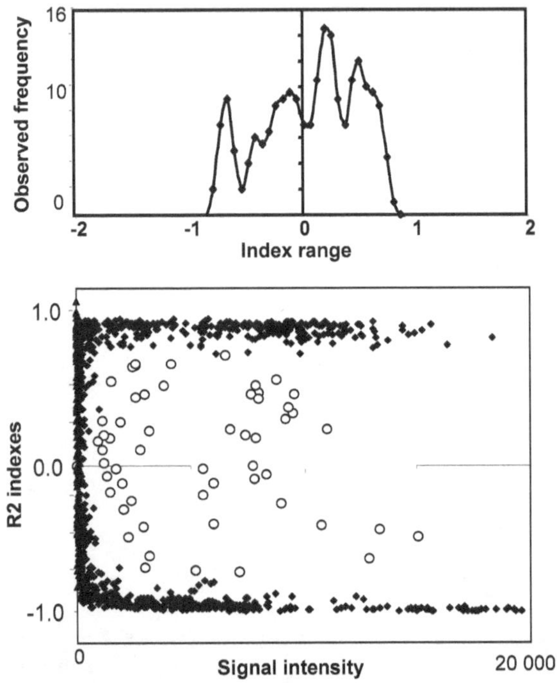

Fig. 5. Top, frequency of observation of the different classes of discrimination index values (mode 2, "abnormal spot" population). Irregular frequency distribution covering the [–1, +1] interval without defined mode is observed. Bottom, observed values of discrimination indexes (mode 2) as a function of signal intensity (sum of green and red channels intensities). The full boxes represent the "normal" (top and bottom) and the "low signal" (left) population and the open circle the "abnormal" population. The figures are shown for optimal values of discrimination threshold (DT = 0.75) and minimal distance threshold (MST = 500) parameters.

3.2.5. Sequence Signature Determination

Sequence signature describes the probability of finding each type of sequence association in the library and the cross-correlation between these probabilities along the sequence. The sequence signature can be calculated both from binary sequence data (macroarray analysis) or probabilistic data (microarray-based sequence mapping). Calculation is straightforward using the principles and formulae described *(2)* and is fully implemented in the "Signa_Process" file template.

1. Paste chip data on the "raw data" sheet and go to the "results" sheet.
2. Set "Bloc count" (number of probe sets) and "Data count" (number of mosaic sequences probed).

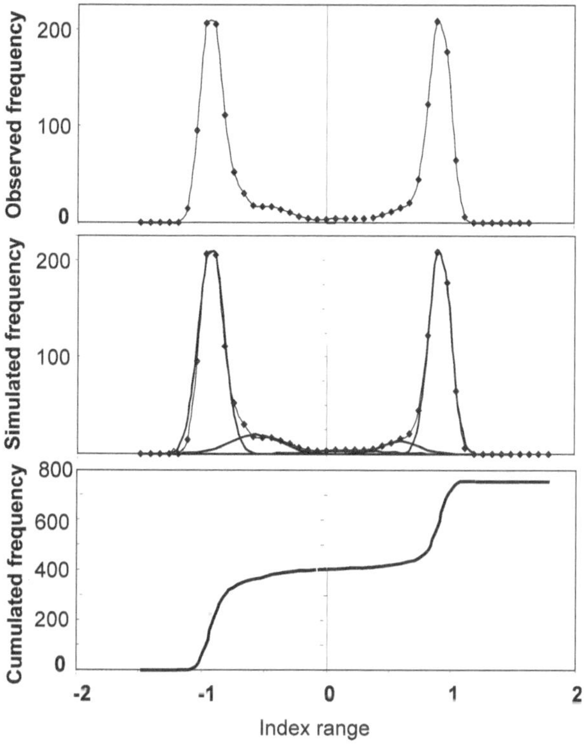

Fig. 6. Top, frequency of observation of the different classes of discrimination index values (mode 2, "normal" population). A bimodal distribution corresponding to the two possible allele types is observed. Center, multi-gaussian decomposition of the top panel distribution. Two major distributions with modes at –0.92 and +0.93 are observed as well as two minor distributions with modes at 0.60 and –0.57. A very limited population with non-interpretable indexes (mode at 0.2 and high dispersion) can nevertheless be detected. This figure corresponds to optimal parameter setting. Bottom, cumulative frequency curve of the previous distributions.

3. Select the considered population (generally "normal") and run the program with the button. Signatures are automatically calculated (*see* **Fig. 7**). The "parental type distribution" panel indicates the parental type proportion for each probed sequence segment. These values are ideally identical for all segments, and equal to 0.5. The "Cross-correlation" and the "Correlation map" panels are two graphic representations of the probability (as cross-correlation coefficients) that two sequence segments in a given mosaic belong to the same parental type. These values are ideally equal to zero except for self-comparisons (value = 1) in the case of a perfectly shuffled library, but the reality is frequently different.

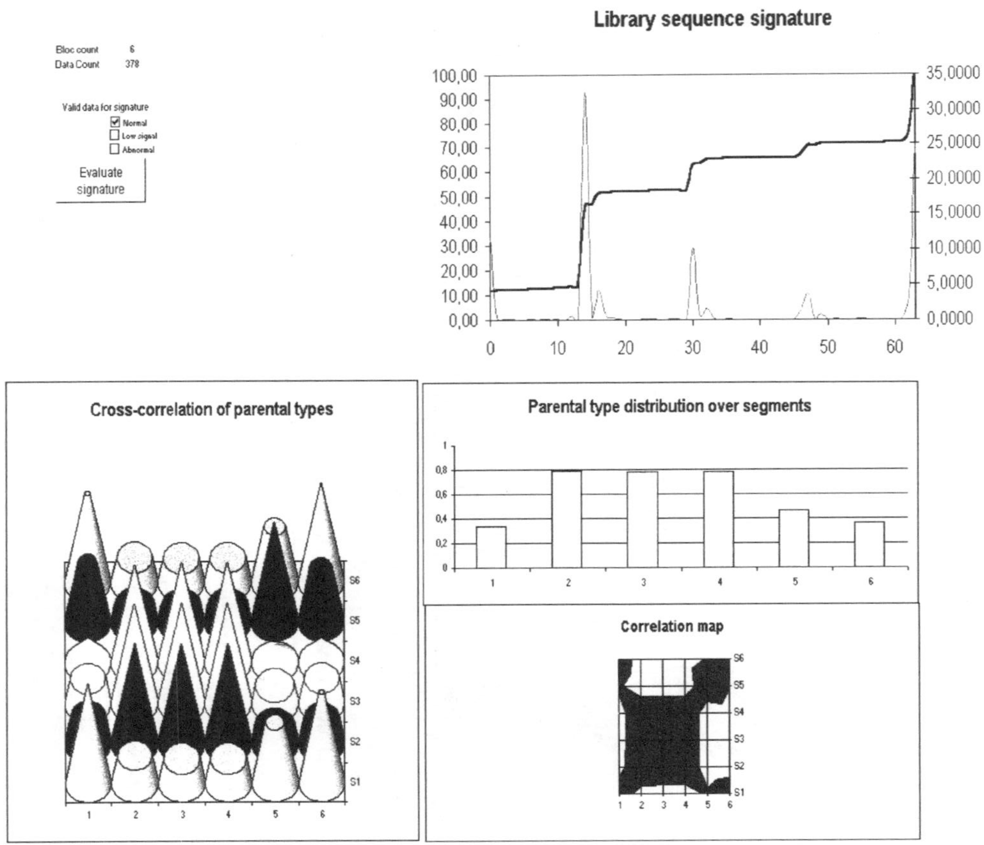

Fig. 7. Graphics interface panel of "Signa_Process" program following microarray based sequence mapping of a poor quality library.

4. The top right panel gives the library sequence signature as absolute and cumulated frequencies. This figure indicates the number of observations of each probed sequence types in the library. An ideal library is characterized by a constant absolute frequency and a linear cumulative frequency curve. In the example, the library was probed with six probe pairs (64 classes) and is of poor quality as illustrated by the discrete peaks of frequencies, the steps on the cumulative frequency curve, the presence of a strongly dominant class of sequences, and the null probability for several sequence classes. The example presented is typical of a library with strong bias and low diversity, which did not merit functional investigation and must be rebuilt with improved methods.

4. Notes

1. Excel® file templates must be used with extreme caution as they contain a mix of data, formulae, graphics, and Visual Basic® routines. Macros activation must be allowed upon opening Excel® sheets or the program will not work.
2. "Copy and paste" on the pure data area is allowed. However, be aware that "cut and paste" in the formula area will usually have catastrophic effects and can lead to unpredictable results. Formula rows, including in intermediate calculation sheets, may have to either be deleted or down-copied (Edit menu) to adjust calculation area length to the number of data rows of a particular experiment. Please read and follow specific comments on the sheets to check if this is applicable or not to the particular area of calculation. Graphics properties (left click) may also need adjustment accordingly and very large sets of data may need editing of Visual Basic® table declarations or default parameters.
3. Never overwrite any formula by a data. Always check on example data provided that no catastrophic event occurred following an action and that the procedure you used is correct before treating your own data. Note that Visual Basic routines are sensitive to sheet names, data geometry on sheets, and exact wording of codes. They can be edited only by experienced users following careful examination of formulae and macros, and understanding the calculations.

Reference

1. Abecassis, V., Pompon, D., and Truan, G. (2000) High efficiency family shuffling based on multi-step PCR and in vivo DNA-recombination in yeast: statistical and functional analysis of a combinatorial library between human cytochrome P450 1A1 and 1A2. *Nucleic Acids Res.* **28,** E88.

Index

f: figure
t: table

G

H

I

K–L

M

FSC
www.fsc.org
MIX
Papier | Fördert gute Waldnutzung
FSC® C083411

Kolibri 360 GmbH
Ferdinand-Jühlke-Straße 7
99095 Erfurt, Deutschland
produktsicherheit@kolibri360.de